W9-AXC-012

Just before noon on December 1, 1974, a TWA jetliner bucking wind-driven snow slammed into the Blue Ridge Mountains near Washington, D.C. The ninety-two people aboard died instantly. Some were never entirely found.

The story begins at dawn in an Indiana farmhouse and ends sixteen months later, at dusk, in a hushed Virginia courtroom. *Sound of Impact* is about the aftermath of an undeniably avoidable accident, from the last confused minutes in a cockpit to young widows' early-morning nightmares. It is about violent death and corporate alibis, public indifference and shattered lives. Collectively, the dead made headlines. Individually, their names meant little to most of us.

But they were somebody to somebody.

Adam Shaw concentrates on the people, the victims and their legacy, the living still struggling to fathom their loss. Through intimate and sometimes wrenching interviews with those left behind, he offers them a voice with which to share their fury and their sorrow.

The voices come from a row house in Terre Haute, a Mexican restaurant in Las Vegas, FBI headquarters in Washington, eucalyptus-shaded ranches above the Pacific, apartment buildings and trailer parks strung across the Midwest; from red-brick army posts and colleges, solemn lawyers' chambers and rudely partitioned government offices. They tell of a tide of recrimination and grief that ripped across a continent after a cold Thanksgiving Sunday, of accepting, of new beginnings. They are the voices of America in anguish.

They were somebody to somebody.

The dead were men and women who paid their bills and dreamed. Who drank and who abstained, who prayed, who sought absolution or drugged forgetfulness. Some were fighters, some

(Continued on back flap)

SOUND

THE LEGACY OF TWA FLIGHT 514

of IMPACT

Y ADAM SHAW

THE VIKING PRESS   NEW YORK

Copyright © Adam Shaw, 1977
All rights reserved
First published in 1977 by The Viking Press
625 Madison Avenue, New York, N.Y. 10022
Published simultaneously in Canada by
The Macmillan Company of Canada Limited

LIBRARY OF CONGRESS CATALOGING IN PUBLICATION DATA
Shaw, Adam.
Sound of impact.
1. Aeronautics—Accidents—1974.
2. Trans World Airlines, Inc.   I.   Title.
TL553.5.S52   973.924   76-49650
ISBN 0-670-65840-5

Printed in the United States of America

Set in Linotype Electra

TO ALL OF THEM, DEAD AND ALIVE

"Doctors bury their mistakes—Pilots are buried with them."

—AN AIRLINE CAPTAIN

"A disaster? No, that wasn't a disaster. A disaster is two 747s full of doctors and lawyers colliding over Manhattan."

—AN AVIATION INSURANCE EXECUTIVE,
DISCUSSING THE CRASH OF TWA FLIGHT 514

AUTHOR'S NOTE

This is a book of fact. The portrayal of lives and events is based on personal interviews, court transcripts, government and private records. The people in it are real as, with few exceptions, are the names of the victims, their families, their friends, and the places they lived and worked.

PREFACE

A TWA jetliner, flying here from Indianapolis and Columbus, Ohio, in fierce winds and heavy rain, crashed and burned in the Blue Ridge Mountains 47 miles west of Washington on its approach to Dulles Airport yesterday killing all 92 persons aboard.

Flight 514, a three-jet Boeing stretch 727 with 85 passengers and a crew of seven, slammed into the western slope of the mountains 23 miles west of Dulles and four miles south of Rte. 7 in Loudon County at about 11:10 A.M.

The aircraft, which had been diverted from National Airport because of weather, sheared off treetops, struck a rocky outcrop, broke up and caught fire, scattering charred bodies and parts of bodies over an area about the size of two football fields.

—The Washington Post
DECEMBER 2, 1974

They were not martyrs, not the victims of a terrorist attack, bomb, or hijacking and, though they died violently enough— all ninety-two of them—it was not in Indochina, Belfast, or the Middle East.

They were the casualties of an accident of our times. Common, devastating—and predictable.

I followed their ghosts, tracked them from the ridge where it all began, back through the cities and villages of their birth—back into the memories of their kin and, sometimes, of strangers. I did this not because they frightened us or clamored for our attention but because they did no such thing. Unwitnessed, they were killed and made no demands.

Farmers, businessmen, government workers, soldiers, housewives, students, cops, athletes: the troubled, the sane, the happy. They were Americans. They were any and all of us.

This book is about picking up the pieces, about alibis, guilt, atonement, about what remains after a huge jet airplane crashes and disintegrates—from the tinsel of what used to be a fuselage to the early morning nightmares dreamt by young widows. It is about victims, dead and alive.

It is also about the legacy of a random wrong—senseless, avoidable deaths; a legacy of slaughter. They died too quietly, anonymously, without time for a whispered curse or prayer, and instantly became statistics in distant computers.

I have seen the tears, asked those left behind to talk, tried to give them a voice with which to share their grief, love, and shame, and their hope. These are ordinary people. Some spoke haltingly, some in a great anguished flood, some said they would rather not talk at all. The living in this book talked because they wished to do so. No effort was made, with the friends and families of the dead, to pry beyond what was willingly offered. This is a book of recollections and reportage, a diary of loss and courage.

CONTENTS

SOUND OF IMPACT

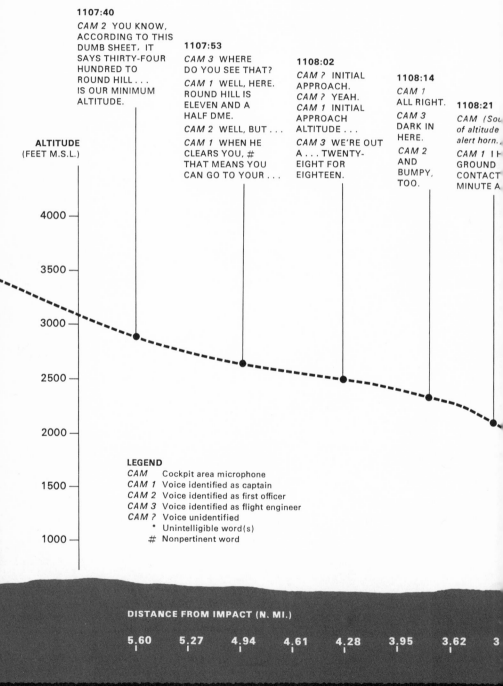

ALTITUDE
(FEET M.S.L.)

1107:40
CAM 2 YOU KNOW,
ACCORDING TO THIS
DUMB SHEET, IT
SAYS THIRTY-FOUR
HUNDRED TO
ROUND HILL . . .
IS OUR MINIMUM
ALTITUDE.

1107:53
CAM 3 WHERE
DO YOU SEE THAT?
CAM 1 WELL, HERE.
ROUND HILL IS
ELEVEN AND A
HALF DME.
CAM 2 WELL, BUT . . .
CAM 1 WHEN HE
CLEARS YOU, #
THAT MEANS YOU
CAN GO TO YOUR . . .

1108:02
CAM ? INITIAL
APPROACH.
CAM ? YEAH.
CAM 1 INITIAL
APPROACH
ALTITUDE . . .
CAM 3 WE'RE OUT
A . . . TWENTY-
EIGHT FOR
EIGHTEEN.

1108:14
CAM 1
ALL RIGHT.
CAM 3
DARK IN
HERE.
CAM 2
AND
BUMPY,
TOO.

1108:21
CAM (Sou
of altitude
alert horn.
CAM 1 I H
GROUND
CONTACT
MINUTE A

4000 —

3500 —

3000 —

2500 —

2000 —

1500 —

1000 —

LEGEND
CAM   Cockpit area microphone
CAM 1   Voice identified as captain
CAM 2   Voice identified as first officer
CAM 3   Voice identified as flight engineer
CAM ?   Voice unidentified
   *   Unintelligible word(s)
   #   Nonpertinent word

DISTANCE FROM IMPACT (N. MI.)

5.60   5.27   4.94   4.61   4.28   3.95   3.62   3

**NATIONAL
TRANSPORTATION
SAFETY BOARD**
Washington, D.C.

DESCENT PROFILE

(REDRAWN)
B727-231, N54328, FLIGHT 514
BERRYVILLE, VIRGINIA
DECEMBER 1, 1974

)8:29
M 2 POWER ON
IS #
M 1 YOU GOT A
H SINK RATE.
M 3 WE'RE RIGHT
ERE, WE'RE ON
URSE.
M 1 YOU OUGHT TO
GROUND OUTSIDE IN
ST A MINUTE.

1108:57
*CAM (Sound of
altitude alert.)*
CAM 2 BOY, IT
WAS . . . WANTED
TO GO RIGHT
DOWN THROUGH
THERE, MAN.
CAM 2 MUST HAVE
HAD A # OF A
DOWN DRAFT.

1109:14
*CAM (Radio
altimeter
warning horn
sounds,
then stops.)*

1109:20
CAM 1 GET SOME
POWER ON. *(Radio
altimeter warning
horn sounds,
then stops.)*

1109:22
SOUND OF

IMPACT

96   2.63   2.30   1.97   1.65   1.32   0.99   0.66   0.33   00

THE FLIGHT

James Michael Davis ties on his blue sneakers. The Clarksburg, Indiana, farmhouse on the hill is cold and in the barn below his father's prize hogs are still asleep. The boy thinks of Virginia and his mother. To the west in Terre Haute, Gerald Seymour tries to keep his dogtags from jingling as he dresses. His brother wakes anyway. "You going?" Across town, Gary Mahmoussian tells his in-laws that he will be back for good by Christmas.

Bloomington. Jake Applewhite packs all his pipes but one, closes his banjo case. His wife dresses their son. In Indianapolis to the north, Tim Harlan wishes aloud that he could stay and play in the new snow with his daughter. "Stay then," says Ellen, his wife. Down in Speedway, Dave Pierce hollers for his mother's tape measure. His freshly laundered uniform shirt hangs from the closet door and Dave, perched on a chair in his skivvies, is trying to repin his second lieutenant's bars exactly where they belong. He measures the collar, pins in the bars, draws back, squints through his glasses. Satisfied, he dresses, collects his loose change and tiptoes into his younger brother's room to stack the coins on the nightstand.

Turning away from a window in their Vinewood Drive

apartment, Mildred DePew tells her husband, "Honey, the weather's awful out there. Why don't you call up and see if they're flying?"

John DePew calls the airport. "They're flying."

The operator at the nearby Hilton brightly announces the day to Richard Brock in room 139, Leonard Kresheck in room 137, and the junior man, Tom Safranek, in room 141. The sky is a swirling band of gray over the hotel's south wing. Kresheck and Safranek breakfast together in the jammed coffee shop. Later, joined by their captain, they gather in the lobby to wait for the bus—three slightly rumpled black uniforms in a sea of double-knits. Brock chats pleasantly with an old Air Force buddy, a businessman, who keeps glancing at the four stripes of braid on his friend's sleeve.

Five hundred and fifty miles away, just outside Washington, D.C., Merle Dameron signs in for work at Dulles International Airport's control tower.

Captain Brock and First Officer Kresheck do the same at TWA Flight Operations at Weir Cook Airport in Indianapolis. Flight Engineer Safranek circles the red, white, and silver Boeing 727 they flew in from Los Angeles the night before. He checks the tires, the landing gear, looks for loose rivets. The plane seems airworthy. He climbs the forward ramp and enters the gray, stale-smelling cockpit.

Echoing those of Indiana, telephones and alarm clocks ring in staggered succession across Ohio to the east. Dogs scratch to be let out into frozen yards; coffee cups and promises are traded over kitchen tables; the snowball fights and long suppertime talks of Thanksgiving fade. It is Sunday now, a miserable wintry Sunday, and the snow—like a turkey carcass— is merely a nuisance to be dealt with.

Sheila Regan tugs on her black cowboy boots, the kind with pointed toes, and adjusts her .38 caliber FBI-issue revolver in its holster so that it fits snugly under her left armpit. Her guard detail at the Mechanicsburg farm has ended but the gun, like the boots, always travels with her. In another farmhouse, a real one in Springfield twenty miles away, Rod Steele slaps dried mud off the soles of his running shoes and, soles up, packs them on top of the newly mended shorts and tee shirt he plans to wear in the Northern Virginia Cross-Country Championships later in the day.

Columbus. Five soldiers—among them a retired brigadier-general and a seventeen-year-old buck private—pack as well. Sitting silently in the kitchen of the small house near the corner of Seigman Avenue, Bob Murday worries about his wife. The first telephone call of the day had been grim, and though Jean tries hard not to show it, she is obviously upset. Murday struggles for words.

Liz Wright lugs her bag down to the front door.
"Breakfast?"
"No," Liz says, "no thanks. Just tea."
Jo Wright suppresses a smile—her daughter really is sticking to her diet—and calls the highway patrol for road conditions. Below their house, the town of Lancaster is blanketed in white.

Laura Meredith and Ruth Speese laugh off an old friend's suggestion that they take out half a million dollars' worth of flight insurance. Doc, who trudged over to their house in Delaware, Ohio, to help pack the car, persists. The two elderly ladies put up a solid front and tell him to stop being foolish.

Up unplowed Route 54 from Clarksburg, down 52 from Lafayette, south on 31 from Peru, east on I-70 from Terre Haute; the headlights poke through the dawn, pale white eyes searching out a common destination. The Indiana passengers of what is to be TWA Flight 514 are gathering.

Eddie Knartzer walks across the airport parking lot with

his father. "I'd rather take off in this kind of weather than land in it."

Wing lights dim in the drizzle, the long jet splashes through the tarmac puddles and, at 8:53 A.M., thunders down the northeast/southwest runway in a cloud of vaporized water, is airborne.

Stewardess Jen VanFossen makes coffee and pours orange juice into plastic cups in the galley behind the flight deck; Denise Stander, five days shy of her twenty-first birthday, passes them around to the eight first-class passengers.

In the antiseptic government cafeteria below the Dulles tower, air traffic controller Merle Dameron orders sausage and eggs for breakfast.

Flight 514 lands in Columbus at 9:32 A.M. and taxis to gate B-1. Six passengers disembark, thirty-nine stay aboard surrounded by piped-in music. The captain and first officer deplane to collect fresh weather information, the flight engineer to hunt for a telephone.

SOHIO fuel truck 9332 pulls up and its driver pumps seventy-two hundred pounds of kerosene into the 727's wing tanks.

From memory Tom Safranek dials a number. Surprised but pleased to hear from his son, Bill Safranek asks where he is.

"Columbus. At the airport." No, there's no time to come home—this is just a fifty-minute layover en route to Washington.

The elder Safranek thinks his son sounds tired and says so. Tom explains: He is on his second hopscotching trip across the country in five days . . . he had spent Thanksgiving in a motel at Newark Airport . . . No, not much fun . . . and he is about to exceed his weekly maximum of thirty hours of flying time. But when, earlier in the morning, he called TWA Operations in New York asking to be pulled from the flight, they just told him he could fly home to Los Angeles as a passenger—once he got to Washington, not before. By regulations, he ought no longer be in the cockpit.

Flight 514 takes off at 10:24 A.M. Eighty-five passengers and seven crew are now aboard.

"Uh, we just got a . . . Cleveland ATC gave us word from Washington, they're apparently making no landings at Washington National—they're landing out at Dulles on runway twelve."* The plane has reached its cruising altitude of 29,000 feet and Safranek is on the company radio frequency talking to the dispatcher in New York. "Wonder if you want us to proceed and go right on in to Dulles . . . or, we've got about forty minutes of holding fuel before we have to divert?"

The dispatcher hesitates, weighing the day's schedule in his head. "Let's try to hold for Washington National," he says. "Uh, if we're unable, we go over to Dulles."

"Okay," says Safranek, "we'll hold a while."

"We just thought," Safranek radios the dispatcher, "maybe you might want to get the wheels in motion, get the people out there, because apparently the winds are on the increase rather than the decrease down there at Washington National . . ."

The dispatcher relents and tells them to plan on "terminating" at Dulles and to "uh, sign it New York dispatch, Epp. E, P, P, at fifteen-forty Greenwich."

---

* This and all subsequent cockpit conversation was taken verbatim—with the exception of the expletives which were deleted—from the National Transportation Safety Board's transcript of the dialogue registered by Flight 514's cockpit voice recorder. Under an agreement with the FAA and the airlines, the NTSB is not at liberty to allow outsiders to listen to the tape, for fear that they might ignore the prurient deletions in its own version and quote the dialogue in full, including any profanity—which, for the purpose of continuity was added, in mildest possible form, by the author.

The author was able, however, to listen to some of the CVR tape, including the final exchange in the cockpit.

"Right on top of it like Dispatch normally is," mumbles Kresheck, the copilot.

Back in coach, United States Marine Corps Second Lieutenant Dave Pierce, conspicuous among the other young soldiers on the flight because he was in uniform, had worried that he might get booted from the plane during the Columbus layover because of his standby status. However, there had been a number of cancellations and no-shows. He planned to be back on base at Quantico for a good night's studying.

Liz Wright's car ride back to school in Virginia had fallen through at the last minute and her parents had urged her to stay in Lancaster at least an extra day to let the weather improve. But she had insisted on leaving, scorning their suggestion that she compromise and take the bus. There were play rehearsals and examinations coming up at Marymount College. Getting out of the driveway had been a touchy, four-try affair, but her father had succeeded and was doubly a hero for having secured her a seat on Flight 514 without a reservation.

Mary Ellen Brogan, one of nine Health, Education, and Welfare (HEW) employees scheduled for a week-long Washington seminar, was older than most of her colleagues but that, in a way, only added to the satisfaction of being included. After twenty years of plodding clerical government jobs, and after months of night school and quizzes, she had recently been promoted to a managerial position.

Weighing the appeal of returning to Representative Andy Jacobs' familiar staff on the Hill against his recent but growing commitment to Atlanta's Andy Young (the New South's champion of an old vision), Jake Applewhite had decided, even before flying home to Bloomington with his wife Susan and Benjy their three-year-old son, that he had to refuse his friend and former boss's offer of a job. Jacobs had understood and respected his ex-legislative assistant's stand and, when they met for dinner Thursday, the matter was not discussed.

It had been a fine Thanksgiving at the Jacobs' house; Jake played his guitar, and Benjy, as usual, charmed everyone.

Late Saturday night—last night—David Holdhusen had asked a girl named Nancy to marry him and she had said yes. But now, along with his Navy buddy and weekend host Dan Hartley, it was back to missile training school in Damneck, Virginia for the young seaman from Aberdeen, South Dakota. Christmas though, was only three weeks off.

Bob Noxon had contracted for additional life insurance and bought his daughter Alice new snow tires for her car over a holiday marred only by the failure of Alice's beau to appear for Thanksgiving dinner. Staunchly, she had ignored the embarrassment, cooked up a feast, and, along with her mother, Laura, now flying back with Bob, seemed genuinely interested in the family tree Noxon was piecing together.

In the cockpit Kresheck complains about the rough ride the automatic pilot is giving them as he pulls out the approach plate for Dulles Airport. He studies it briefly and asks Brock for the wind forecast.

"Well, uh ... shit, twenty-five gusting to forty-three," Brock says. It's going to be a bumpy descent.

Safranek reminds his captain that his breakfast tray, brought forward by Jen VanFossen a few minutes earlier, is cluttering up his panel. Brock asks him to pass it up. Kresheck says, "I got her under control, boss, I guess."

"Have you?" Brock jokes. "That's good, 'cause I ain't got the mother." He takes the tray from Safranek.

"Okay, I'm flying it."

"You flying it," Brock tells his copilot.

The plane jolts.

"Damn," says Brock trying to settle back in his seat, "he's gonna get me all bouncing around here ..." The Captain considers his eggs. "Aw look at that sick-looking shit, look at that sick-looking crap! I think I'm going to have time to eat at Dulles."

The stewardesses pour last refills on coffee as the plane sinks toward 23,000 feet.

The unexpected gift of a ticket home for Thanksgiving had made the prospects of Tuesday night easier for Cathleen Casey to contemplate. She tried to believe Tom Skinner's promise that it wouldn't be so bad. "You can see him, but he can't see you. Really." Still, Cathleen dreaded seeing that face again, the face with the gun, the eyes, the voice.

"Give me your purse, ma'am!"

The hysterical politeness of it all . . . ma'am. And the old-fashioned rabbit ears of the sawed-off shotgun snapping down. And the frozen silence broken by a scream, a wild, disembodied animal shriek. The cold, truncated barrel at her forehead. And once more, the dull, metallic trigger click.

She had run, unharmed, somehow, from her little white and orange Pinto. Then, two days later, Detective Skinner's reassuring, sympathetic Washington cop face was staring at her across her apartment living room, listening. "I need to get away from this city, I never dreamed it could happen to me . . . But I can't afford to go home." And Skinner's handing over the folder full of mug shots. Then the face jumping out at her like a photograph of a snake in a magazine. Skinner's pro forma question—"Are you sure? On a scale of one to ten, how sure?"

Her answer. "Nine and a quarter."

Skinner, grateful now, saying, "Thanks, Cathleen. I'll be in touch."

He had called the day before Thanksgiving. Excitedly, she told him her father had bought her a ticket out to Columbus. Skinner had said that was nice and that he thought they had her man, Hector Diaz, nineteen, a street kid. There would be a lineup. Tuesday at seven.

She said she didn't really want to go. Skinner explained that an indictment would be unlikely without positive identification. As a lawyer, Cathleen knew he was right. And now she was coming back to face Diaz, who had spent his holiday weekend getting arrested and pacing the District Court's central cellblock waiting for a hearing, wondering which rap they'd get him on this time.

Up front Brock asks whether anyone has told the passengers about the rerouting.

"I'm gonna do that," Safranek says.

"That'll make everybody happy," quips Kresheck.

The flight engineer begins a mock announcement. "Ladies and gentlemen, for those of you who live near Dulles . . . for those of you who live near Dulles, we got good words for you—"

"The good news is," Kresheck interrupts, "if you live near Dulles, we're gonna land there. The bad news is if you live near National, you gonna be on at Dulles."

"Yeah," adds Safranek, "you got a bus ride coming."

The captain finally makes the real announcement. Groans can be heard from the cabins behind them. National Airport, on the banks of the Potomac, is only a fifteen-minute cab ride from downtown Washington; Dulles, in Chantilly, Virginia, is forty-five minutes and an extra fifteen-dollar hack fare farther out.

Of the dozens of friends and relatives waiting on the ground, at the wrong airport, for the passengers of Flight 514, only Mario Mathieu knows where the plane—with his wife aboard—is actually headed. A twenty-four-year-old TWA baggage handler at National, Mathieu has just returned from TWA Operations where his friend Joe Hernandez let him listen in on radio transmissions between the 727 and the tower.

10:52 A.M. Mathieu watches television in the ground crew lounge (National Airport is closed; there are no planes to unload—an easy day) and ground control has just directed Captain Richard Brock to cross the three hundred degree radial twenty-five nautical miles northwest of Dulles Airport at eight thousand feet.

Brock repeats the instructions and Kresheck acknowledges them. Pulling out the Dulles approach plate, the captain says,

"Okay, let's see what that VOR has to offer... It's pretty much right down the runway."

"Yeah," agrees Kresheck.

Brock tells him their approach speed ought to be 127 knots. The plane bucks.

"Hey, hold still."

"Yeah." Kresheck levels it off.

Brock tabulates the approach speed and says, "Well, this is a bunch of shit too, 'cause on a go-around, you got to use what you need, you know, if there's a building coming up..."

A brief exchange between the tower and a Braniff flight asking for an update on local weather crackles over the cockpit radio.

"They're still trying to keep him out of the river," snorts Brock.

Kresheck concentrates on keeping the jet on course as Safranek reads out the preliminary landing check list. The two senior men answer, flipping switches, turning knobs. Fourth on the list is lighting the spotlight at the base of the plane's tail illuminating the big red TWA logo above.

Kresheck makes a small course correction, lines up the 727's nose on the VOR [visual oral range] radio beacon extending its signal out from runway #12. "You know," he muses, "they get this [bad weather] once in a while. We went in out here at Dulles one time before. We diverted out of National, went into Dulles—same thing. The wind was blowing like a son of a bitch—it just beat the hell out of us going down there."

But he's nonchalant, breezily complaining only that he hasn't landed there in so long he wouldn't even know how to find the office where crews wait for the bus to town.

Brock and Safranek get out their approach plates. The captain is concerned with the weather and with landing on an unfamiliar strip. Earlier in the flight he joked that he hadn't been in to Dulles "in a hundred years." The approach plate, a rectangular sheet of thick white paper crisscrossed with lines and numbers indicating air routes and minimum altitudes for those routes within a twenty-five-mile radius of the airfield, is their only guide.

Looking at his copy of the plate, the copilot notes that "You gotta taxi around here clockwise."

"Don't worry about that now," says Brock.

"... Well, we have to ... you have to go around there c-clockwise ..."

Brock politely cuts him off. "Okay. You keep me posted."

"It's pretty easy the way we're coming in," Kresheck says.

The plane crosses the 11,000-foot mark, sinks toward turbulent air.

The crew calls down to the Leesburg Air Traffic Control Center. "Center, TWA five fourteen—do you paint anything uh, any obstruction between us and the airport as far as weather?"

"Ah, no sir," the Leesburg controller replies. "I'm not showing any significant precipitation along your route of flight. It's mostly to the south of Dulles and the southeast."

"Okay, we're not painting anything either. We just wanted to make sure."

Fifty-six seconds later the Leesburg controller is back on the air. "TWA five fourteen descend and maintain seven thousand [feet] and contact Dulles Approach on one one niner point two—so long."

Brock calls down to the Dulles tower to say his plane is heading down.

Merle Dameron acknowledges. "TWA five fourteen, Dulles, roger. Proceed inbound to Armel, expect VOR DME approach runway one two."

The crew listens in as Dameron vectors two other flights toward the same runway and gives them a rundown on conditions on the ground. "Dulles weather measured ceiling nine hundred [feet] overcast, visibility five, light rain, wind is zero nine zero degrees variable at three zero knots and the altimeter [setting] two niner seven zero."

"Gusting to three six," Brock mumbles to himself.

11:04.16 A.M. Dameron is back on the radio. "TWA five

fourteen, you're cleared for a VOR DME approach to runway one two."

Brock reads back the clearance and, again, looks at his approach plate. "Eighteen hundred's the bottom," he says. It is 11:04.27 on the morning of December 1, 1974, and Richard Brock has just made the biggest mistake of his nineteen-year career with TWA.

"Start down," says Kresheck.

"We're out here quite a ways," cautions Safranek. "I better turn the heat down."

"Yeah," agrees Kresheck, "that's pretty high."

"Huh? Oh yeah," Brock says vaguely, still thinking about the approach. "Okay . . . yeah, well, I—I meant to do that, I just looked at that and he gave me this clearance. Then I saw the windshield wipers clear all the ice off . . ."

The two junior men in the cockpit remain silent, allow their captain to verbally compensate for his momentary inattention about the de-icing systems.

"That's good, Tom," Brock fumbles on, "that's keeping your eyes open."

The altitude alert horn on the barometric altimeter sounds, warning the crew that they are at five thousand feet.

"Keep me out of trouble, fellow," says Brock.

Kresheck grumbles about the instrument needles jumping around; at least the Air Force jets' gauges were rubber-mounted to keep them stable. Flight 514 is in the clouds now and getting a rough ride.

From the rear of the cockpit Safranek teases Brock. "Pat, I thought you said the second day was an easy day?"

"Yeah . . ." chips in Kresheck.

"Oh it is," the captain replies. "Hell, I'm not working hard."

"I'll check back with you in an hour or so," Safranek says lightly. He and the copilot discuss the wind, agree it is rough but, as Kresheck notes, "not that you can do anything about it."

Brock interrupts them; he'd been bothered by the approach plate and now realizes why.

"You know," he says, "according to this dumb sheet it says thirty-four hundred [feet] to Round Hill . . . is our minimum altitude."

Three minutes and twenty-one seconds earlier, when he received Dameron's clearance, Brock had said eighteen hundred feet, not thirty-four hundred, was "the bottom," the safe approach altitude. The jet is already below 3400, but not by much.

Safranek asks, "Where do you see that?"

"Well, here," Brock says, pointing to the numbers on the far left side of the sheet.

Kresheck glances at it. "Well, but . . ."

"When he clears you," Brock says, "hell, that means you can go to your . . ."

"initial approach . . ."

"Yeah."

"initial approach altitude," concludes Brock.

Neither the copilot nor the flight engineer contradict him. All three are either agreed, or silence whatever remaining doubts they have about Dameron's clearance.

"We're out a . . . twenty-eight for eighteen," Safranek says.

"One to go."

It is Kresheck's turn to be puzzled. "How come that fucker didn't light up?" He expected the amber light on the altimeter to go on, indicating the plane had passed through the 2800-foot mark.

"That's 'cause we have a thousand to go," says Brock.

"Well, we had the wrong numbers in it."

The flight will soon be over. "All right," Brock says.

The four-and-a-half-year-old plane shudders through the cloud layer, snow and hail alternately streaking the cockpit windows. The seat belt sign is on; even the stewardesses are strapped in.

"Dark in here," Safranek says.

"And bumpy too," agrees Kresheck.

The altitude alert horn sounds again. They are at 1800 feet.

"I had ground contact a minute ago." Brock has just spotted the steely gray waters of the Shenandoah below and slightly ahead. Kresheck looks out—"Yeah, I did too"— glances down at his altimeter, sees they are slightly below eighteen hundred. "Get some power on this mother."

"Yeah," says Brock, "you got a high sink rate."

They are over the river.

"We're going uphill."

Safranek: "We're right there, we're on course."

"Yeah," agree Brock and Kresheck.

Brock: "You ought to see ground outside in just a minute."

The plane lurches, Kresheck fights for control.

"Hang in there, boy," mutters Brock.

Safranek: "We're getting seasick."

Again, the jet dips. "Boy," Kresheck exhales, "it was . . . it wanted to go right down through there, man! Must have been a hell of a down draft."

The radio altimeter horn sounds, then stops. It is preset to go off at five hundred feet. The broad, level plain leading to runway 12 should have long since come into view. Instead, the horn is triggered by the foothills of the Blue Ridge, hard ground building up into a deadly wall ahead.

Six seconds later Brock says, "Get some power on."

Leonard Kresheck reaches for the three throttles to his left.

The horn again.

Kresheck has the throttles.

At 198 knots the screaming jet, left wing slightly dipped, hits the first black oak poking seventy feet up through the mist of Weather Mountain. It is 11:09:22 A.M.

Within the next fraction of a second, like a giant lawn mower out of the sky, the plane cuts a 380-foot swath through pine, spruce, and oak, ripping and tearing its way lower and lower toward Route 601.

The men in the cockpit are already dead—speared by tree limbs shattering through the windshield. The jet crosses the road at fifteen feet, sinking. Nose-first, it bludgeons into the basalt ledge on the far side of Route 601 and, quite simply, ceases to exist. A bloody rain of limbs and debris first rises and then falls through the bare branches. Fires spread on the rock, on the soggy forest loam.

The fog and the silence roll back in.

PROPER IDENTIFICATION 2

The tidy data block representing the jet's position and speed no longer figured on Merle Dameron's radar screen. The controller called up. "TWA five fourteen say your altitude."

The plane had crashed thirty-two seconds earlier.

And as it did, the lights went out in Jack York's house up on Weather Mountain. An even-tempered man, the Loudon County farm agent patiently laid down his trowel and bricks, rose from his knees, and groped his way upstairs from the basement where he had been building a fireplace.

Dameron waited seven seconds before calling again. "Trans World five fourteen, Dulles."

Suspecting that the brutal storm raging outside had knocked down the exposed power line strung along Route 601, York slipped into a rain slicker and walked to his pickup truck.

Waiting another eleven seconds, Dameron—a slight edge now in his voice—asked, "TWA five fourteen, Dulles approach. Do you hear me?"

York drove north on Route 601—a narrow blacktop paralleling the ridge line and the Appalachian trail—found nothing wrong with the power line, turned around and backtracked.

"TWA five fourteen, Dulles, one two three, three two one, how do you hear me?"

Passing his house, York continued south on Route 601.

Dameron called his shift supervisor and radioed up to another flight. "TWA ninety-nine would you give TWA five fourteen a call?" The two men listened. Nothing.

Through the slapping windshield wipers, York found what he was looking for. Coiled in the road a thick black power cable fizzled in the rain. York got out of his truck and peered through the mist and drizzle. To his left, the forest floor was on fire. "A tanker's blown up," he thought, racing for his truck. From home he called the local fire board to report what he had seen. The fireman on duty down in Leesburg told him a plane, a "seven-something-seven" was suspected down on the mountain.

"TWA five fourteen from TWA ninety-nine, do you read?" The messages echoed through the sky. Dameron, relieved of his position, stood by.

"That's it," said York.

By noon, the Virginia State Police had blocked off Route 601 down at Snicker's Gap where it joined the main road crossing the ridge between the Shenandoah Valley and the coastal plain. Up at what was already known as "the site," a continuous stream of volunteer firemen, rescue workers, and policemen poured in, parked their cars and trucks, stepped out into the rain and smelled death. Oliver Dube, the Loudon County fire marshal, was one of them. Walking toward the rock ledge which gutted the plane, he was met by Bob Reily, captain of the Round Hill rescue squad.

"Well?"

Reily shook his head. "We've got one hell of a goddamned mess."

"Look like anybody make it?"

"No," Reily said, "I doubt it."

Dr. George Hocker, the county medical examiner arrived a

little later; he'd been sleeping late, woke when the local hospital called to say they were on alert. Deputy Sheriff Alvin Lynn picked Hocker up in a squad car and together they rode to the mountain. On the way they heard over the radio that there were no survivors. Hocker's medical services would not be needed. A far grimmer task awaited him; as coroner (a job for which he had volunteered years earlier, and which normally consisted of signing death certificates and of autopsying the occasional drunk) he would share the responsibility for identifying the ninety-two victims of TWA Flight 514.

Police and so-called "rescue" workers were already fanning out through the wreckage when Hocker got to the crash site. Immediately he ordered that, as they were gathered, the men "try to keep the pieces of bodies together." It was not easy. The plane had hit the first tree at roughly 228 miles an hour. In less than a second, it had not only stopped but disintegrated, flinging passengers—some still strapped to their seats— through the dense forest and onto the shrapnel-like remains of the aircraft.

Dube later recalled that "some of the bodies didn't have any clothes on. I don't remember seeing any with shoes on. There were pieces of arms and legs wrapped around trees. I saw one leg embedded in a tree . . . so many dead people at one time . . . How can ninety-two people die at one time? A plane crash. That's how."

Hocker was reminded of a Vincent Price movie. Thankfully, he was to say, "the rain has a habit of cleaning bodies up, but you got the impression that someone had put all these people in a bag and just shook them out."

"There were pieces all over the mountain," said Bill Peters of the Sterling Park rescue squad, "and they weren't very big pieces."

To Brett Phillips, a young editor on the Loudon *Times-Mirror*, the victims looked exactly like "football dummies or thirty-year-old Raggedy Anns." With the sweet, persistent smell of death clinging to his clothes and to the back of his throat, Phillips drove down from the mountain and wrote a story for his paper which began, "The grim spectacle that unfolded on the Blue Ridge Sunday seemed to have only one consistency—an aura so macabre that it approached the unreal.... Standing on the running board of a fire truck, and looking down at his grim audience, the medical examiner assumed the appearance of some sort of high priest, looming above his congregation in the fog to deliver a sermon on death. But it had to be done that way, and it was. With darkness threatening the mountain, the newly briefed searchers moved off into the gloom, looking like so many soldiers carrying their stretchers into battle rather than out of it."

As the men in the yellow, black, white, and red slickers poked through the aluminum tinsel of what used to be an aircraft, some stumbling off behind the mangled trees to retch, Mario Mathieu watched *Star Trek* on television, munching on a bologna sandwich in the TWA personnel lounge at National Airport.

"I saw no bulletins," he was to say, "everybody else said they saw bulletins. Then this friend of mine walked in and said, 'I'm gonna tell you something nobody is supposed to know. I just read it on the teletype. Flight five fourteen just crashed twenty-five miles from Dulles.'

"I said, 'What?' and my heart stopped. Everybody knew my wife was on that flight and someone said, 'Well, his wife is there,' and Guthrie, my friend, felt really bad because he was the one who told me. And I jumped up and dropped everything, my food and everything else, and ran to Operations where Joe Hernandez is, and I said, 'Is it true?' and he said, 'Yes.' So I grabbed a phone and called [the airport in] Columbus, Ohio, but I couldn't talk so this friend of mine asked if she was on the passenger list, and she was.

"Then they told me my father was upstairs, at the desk. I went out there and we waited. I just couldn't take it, we waited. And some lady said, 'There are survivors, there are quite a few survivors.' I don't know who she was ... some maniac."

Led by Rudy Kapustin, the National Transportation Safety Board's "Go Team" drove to Weather Mountain, each man hoping that the Cockpit Voice Recorder and the Flight Data Recorder would be found intact, that the magnetic tape of one and the stylus scratchings on metal foil of the other would tell them what had happened. At the old FBI building, the fingerprint specialists gathered their kits—plastic surgeons' gloves, scalpels, cardboard cards—and headed west, across the Potomac into Virginia. Dr. Hocker needed help.

A state trooper liberally ticketed cars parked on Route 50 at Snicker's Gap; by the ridge line another trooper stumbled through the shredded trees in a daze. He carried, at arm's length, a holstered .38 revolver and a pair of dented handcuffs he had just found near the mutilated body of a woman wearing black cowboy boots.

In the old school at the Bluemont Community House, Dr. Hocker set up a morgue. He ordered the heat turned off and all the windows opened. The first body bags were carried in.

Alan Clammer in Montecito, California, was on the phone to his mother in Philadelphia when the operator interrupted to tell him he had an emergency call. Clammer, a TWA flight engineer, lived a few miles up the hill from Tom Safranek and his wife, Vickie.

"I hung up the phone," Clammer remembers. "It rang again. It was Vickie and she was hysterical. Mr. Safranek [Tom's father] had called her to say that apparently the plane had gone off radar and (she asked) could I please see what I could find out. Of course I was shocked and said I'd see what I could find out and that I'd be right over. And I told her right then, 'Vickie, I gotta tell you that you've got to expect the worst because you don't just set these planes down in a cow pasture like they did many years ago.'

"I called crew scheduling in L.A. and they gave me the brush-off. They said, 'We're not at liberty to give out information.' 'Listen,' I told them, 'I'm Tom's best friend. I've just gotten a hysterical call from his wife. I'm gonna be with her in five minutes. I want some information.' They said, 'Well, call us back when you get to her house.' I got to her house and Brenda [Clammer's wife] went with me and we comforted Vickie for a few minutes. I called crew schedule back; they put me on 'hold' for an ungodly amount of time. I'm sure they were just seeking official approval to give me the true word, which must have come through, because they got back on the phone and told me there were no survivors. Vickie had gone to the bathroom when I made the call, and after they told me 'no survivors' I just left the phone off the hook because I didn't want anybody else to call in and tell her. Brenda and I walked down the hall towards the bathroom and met her halfway.

" 'Have you found anything out?'

"And I just said, 'Yes. Tom's dead.' Just like that.

" 'I've got to sit down,' she said, and she sat down and cried for a while. Then I put the phone back on the hook and it rang immediately and I picked it up and it was Mr. Safranek, and like a dumb shit, I said, 'Oh, how are you, Mr. Safranek?' How was he supposed to be? I don't know why I said that ... Things you say on the phone all the time ...

"And it seemed to help Vickie a great deal at that time to begin calling people; she wanted to tell people what happened. All she could do for the next few hours was call people. It was emotionally very tough for me to sit there, hearing the same thing over and over again. I loved Tom."

A continent away, in the overheated and crowded main lobby at National Airport in Washington, George Speese, Jr., and his younger brother, John, heard that Flight 514 had been delayed. Forty-five minutes later another announcement asked that the people waiting for the TWA flight gather at the check-in counter.

"I was half asleep," remembers George, "John went to the counter and they announced that the plane had been rerouted to Dulles. Fifty minutes later, they again asked for people waiting for Flight five fourteen to come to the counter. (By now it was past noon; the plane had crashed more than an hour earlier.) I was making a phone call, so John went to the counter. They told him the plane had *landed* at Dulles, and that it would take twenty to forty-five minutes and that [the passengers] would be on buses to National."

Speese and his sixteen-year-old brother were waiting for their aunts, Laura Meredith and Ruth Speese, who were flying in to help celebrate the fortieth wedding anniversary of the boys' grandparents. It was a surprise visit though, and George had to keep running to the phone to explain to his father why the two ladies hadn't yet arrived.

"At that point, John didn't feel so good. I suggested that he go outside, get some air. We'd been sitting for two and a half hours. He went outside and didn't come back. I went to get him. He was vomiting, he was actually physically sick. He said he knew something was wrong. At that point I was still ...I figured TWA had no reason to tell us the plane had landed if it hadn't. So I was just trying to console him, telling him there wasn't anything wrong. But he still kept shaking. Then they asked again for people waiting for Flight 514 to come to the counter. John went to the desk; I went to make a phone call. I don't know whom he talked to, but then they told him the plane had crashed."

"Proper identification is the last thing you can do for them." At the impromptu morgue in the old Bluemont school-house, Dr. Hocker was already at work on Body Bag #1. It was to be labeled Rodney E. Steele—the runner. Four ten-foot tables lined one wall and a fifth stood in the middle of the room; it was the crucial one where dental charts (mailed from dentists' offices and army posts around the country, alerted about the crash by the families of the dead), when they arrived, would be matched against what was left of mouths.

As though partaking in some peculiar ritual of atonement,

volunteers from TWA handled the body bags, brought them down from the mountain and up the schoolhouse stairs. They stacked them against the back wall and, when there was no more room, laid them in a refrigerated meat truck parked outside, its generator running.

The FBI fingerprint specialists were there, dressed in white coveralls, surgical gloves and blue baseball caps with the Bureau's seal on their peaks.

John Walters, a veteran print sleuth, told his men to "pick up as many hands as you can find and start finger-printing them." Not pleasant work, he was to say, but "once you step up and open that first bag and look in, from then on they're all the same."

Bob Wright had just gotten back to Lancaster from Columbus and the airport. He'd mixed himself a martini, slapped together a turkey sandwich and settled down to watch a football game on television. His wife, Jo, an ex-airline stewardess, was cleaning up the post-Thanksgiving mess when Maggie, their youngest daughter asked, "Mother, what plane did Liz take?"

"TWA," said Mrs. Wright. "Why?"

"Never mind."

But Maggie had had a strange look on her face. She'd been listening to the radio. Mrs. Wright went down to the den where her husband had just seen a bulletin on the crash.

"Bob," she said, "call the airport."

As he hunted for the number Bob Wright, a large, gentle bear of a man, told his wife to stay calm. "We don't know that it's her plane."

He got through to TWA. Mrs. Wright started to scream and as she later said, "I just kept screaming." A doctor was called to give her a sedative. He was followed by two priests.

Bob Wright kept telling her there might be survivors. Over and over his wife repeated that it wouldn't matter, that Liz "could never fight her way out," that "she's too little to fight her way out," that "she'd never make it out of that plane!"

None of it made sense. "Why? Why her? Why?"

In TWA's huge New York headquarters at 605 Third Avenue, secretaries hastily summoned to work manned the phones, calling everyone on "the list," from Chairman of the Board Charles C. Tillinghast, Jr., Chief Pilot (and head accident investigator) Bill Sonneman, to Fred Schaffhausen, the head of TWA's account at Associated Aviation Underwriters. At New York's directive, a "hold" signal was punched into the airline's central reservation computer to prevent some well-meaning clerk in Columbus, Indianapolis, or Washington from dialing up the passenger list to confirm or deny a victim's presence on the flight.

At 4:00 P.M., Jack Jones got word from New York that there were no survivors. The TWA manager at Columbus Airport decided that the "most humane" way of breaking the news to the victims' families would be to send teams of TWA employees directly to their houses instead of answering the endlessly ringing telephones. Jones divided his managerial crew into squads of two, gave them each lists of names and addresses, told them to be as tactful as possible, and sent them off through the snow.

Tom A. Ansel, TWA's general manager in Indianapolis, heard of the crash on his car radio as he drove home from a family luncheon in Richmond, Ohio. Pulling off the freeway at the very next exit and into a gas station, he had to wait to use the phone. Through the open door of the public booth he overheard a man beseeching someone at Weir Cook Airport to tell him whether or not a relative of his had been on Flight 514.

"Quite a coincidence," thought Ansel.

Not trusting that Dulles Airport would remain open, Frank Parisi of TWA's New York public relations staff cajoled his way onto a packed Metroliner to Washington, his brief-case filled with statistics. Happy ones—yearly the airline safely flew more than fifteen and a half million passengers through the skies and in seven years had not, until today, had a fatal domestic crash; and the unhappy ones—only three months ago, a TWA Boeing 707 plunged 30,000 feet into the Aegean during a flight from Athens to Rome, killing all eighty-five aboard. But in that incident the cause, apparently, had been sabotage.

Equally skeptical about Dulles weather, Associated Aviation Underwriters' senior man in charge of TWA's account flew only as far as Baltimore before renting a car for the long drive to Route 601 and the accident, now officially designated as the Berryville, Virginia crash. In his briefcase he carried an AAU checkbook. There would be undertakers, hired hands, and supplies to be paid for. The insurance company must come across as Big Daddy, ready to hold hands and foot the bill.

And across the country, during halftimes of professional football games, the news bulletins flashed on television screens; people paused in mid-motion to listen, to hold their breaths, to rush for the nearest phone, dreading confirmation that the someone they knew was on that plane.

By dusk the police, investigators, and rescue workers had retreated from the wreckage, driven off by a lightning storm. Thirty-one closed body bags had been brought in. The rest would have to wait until dawn.

In room 824-A on the eighth floor of the National Transportation Safety Board headquarters in Washington, Paul Turner unscrewed the charred orange outer casing of the Fairchild A-100 Cockpit Voice Recorder recovered earlier in the day from the debris of Flight 514, extracted five screws from the stainless steel inner casing, and reached down for the fireproof cocoon holding the tape. It appeared undamaged. Quickly, Turner made a duplicate, then threaded the original into a big table tape recorder. Curious, helpless—a spectator at a killer fire—he punched the PLAY button.

"Landing to be at runway twelve at Dulles?"

"That's what he said."

"I haven't been to Dulles in, shit, thirty years."

"Nor have I—hundred years."

The voices Turner had just heard but could not yet identify belonged, in succession, to Flight Engineer Thomas Safranek, Captain Richard Brock, First Officer Leonard Kresheck, and Brock again.

Their voices were clear, two of them nasal, vaguely western.

The tape played uninterruptedly for thirty-one minutes and seven seconds. The spool thinned down to its last revolutions. Brock said, "Get some power on."

A pause. The altimeter warning horn sounded mutely, followed by the second-long, insanely violent, wrenching noise of an airplane breaking up—the sound of impact.

THE HEARING

Preceded by nearly two months of semipublic name calling, pilots, controllers, government officials, and TWA representatives finally met face to face on January 27, 1975, as the National Transportation Safety Board began its public hearing on the crash of Trans World Airlines, Inc., Boeing 727-231, N54328.

The vast blue-carpeted meeting room in the Holiday Inn in Crystal City, Virginia, was stuffy and the public address system sputtered; someone had plugged his private tape recorder into the wrong socket. Retired Coast Guard Rear Admiral Louis M. Thayer, the hearing chairman, opened the proceedings by setting ground rules.

"The hearing," he said, "is being held specifically for the purpose of determining the facts, conditions, and circumstances which will assist the Board in determining the probable cause of the accident and ascertain those measures which will best prevent similar accidents in the future. It is to be clearly understood that the Board has not arrived at any conclusions regarding the causal or contributing factors relative to this accident."

Squinting out from behind his sunglasses at the bank of

TWA and government attorneys facing him under the glare of television lights, Thayer paused. The local press had been running stories based alternately on the airline pilots' and air traffic controllers' accusations—in which each, effectively, blamed the other for the crash; Captain Brock versus Mr. Dameron.

"The inquiry," Thayer continued, "is not being held for the purpose of determining the rights and liabilities of private parties, and the Board makes no attempt to do so."

Again he paused. The assembled attorneys nodded politely. For the last six weeks both sides had been meeting privately to try to determine, and hopefully to agree, on where the blame lay and thus avoid the rancor and publicity of a public trial on the issue of liability. Beyond either side's obvious aversion to accept responsibility for the death of ninety-two people, their real fear of being found at fault was generated by the financial aspect of the case. The party most responsible for the crash would bear the brunt of defending scores of damage suits brought against them by the victims' relatives, and neither TWA nor the government (who, through the Federal Aviation Administration, trains and employs air traffic controllers) relished the prospect of paying out settlement claims and jury awards which might well exceed $12 or $13 million.

The meetings, however, had been fruitless, with TWA unofficially sticking to its position—voiced by spokesmen for the Airline Pilots Association—that when Merle Dameron told Flight 514, then 45 miles from the airport, it was "cleared for a VOR DME approach to runway twelve" he had unequivocally led the pilots to believe that they were permitted to descend to their final approach altitude which the Dulles approach plate correctly depicted as being 1800 feet. Furthermore, the pilots hinted, the TWA jet was at that time following an unpublished course—one given them by a controller rather than printed on their charts—and as such was dependent on the controller for information governing altitude

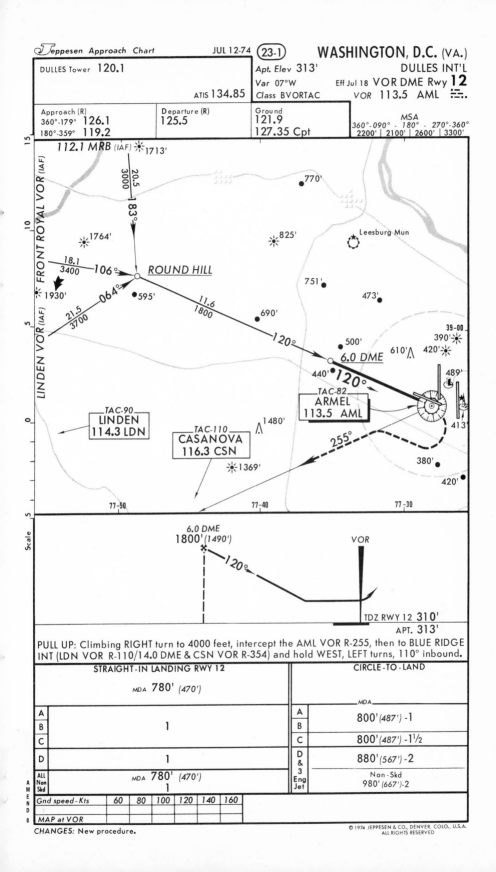

restrictions. Receiving none, it was therefore logical that Brock and his crew felt safe in coming down.

The controllers, through their union, held their own news conferences, acrimoniously contesting the pilots' position by saying that (1) Dameron's clearance had in no way permitted the jet to descend to 1800 feet; (2) that a pilot was ultimately responsible for his own "terrain clearance"; (3) that any pilot worth his wings would know that 1800 feet was far too low to be while still forty-four miles from the runway; (4) that the crew, when confused about the clearance ("You know, according to this dumb sheet it says thirty-four hundred to Round Hill is our minimum altitude . . . Well, but . . . when he clears you that means you can go to your . . . Initial approach . . . yeah . . .") ought to have radioed down for clarifi-

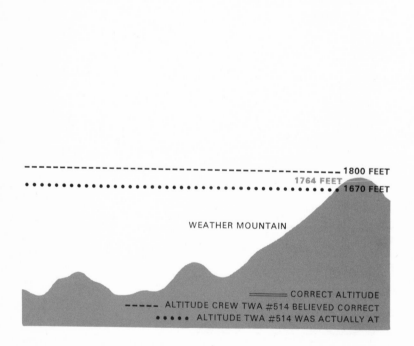

3400 FEET

1800 FEET

1764 FEET

1670 FEET

WEATHER MOUNTAIN

CORRECT ALTITUDE
ALTITUDE CREW TWA #514 BELIEVED CORRECT
ALTITUDE TWA #514 WAS ACTUALLY AT

cation, as had the captain of an American Airlines flight when he was bewildered by a similar clearance; and finally that (5) had the TWA jet actually been at 1800 feet, it still would have cleared the 1764-foot hill. But they had been even lower than that, at 1670 feet.

Responding to public, aviation industry, and congressional outrage over an apparently long-standing and obviously lethal terminological gap between pilots and controllers, Thayer concluded his opening speech by announcing that the inquiry would "extend far beyond the crash of TWA Flight five fourteen [to] include an in-depth investigation of the operation of the Federal Aviation Administration Air Traffic Control System, particularly the formulation and approval of procedures, the issuance of approved clearances, and the interpretation of published procedures." TWA's attorneys politely refrained from looking at their FAA counterparts.

The television cameras whirred and Jack York was called as the hearing's first witness. Shyly he told of his truck ride up and down Route 601. The crowd fidgeted, waiting for Merle Dameron—the one man alive directly involved in the crash, in effect its only survivor. For three and a half days, weather forecasters, radar specialists, air traffic controllers, and pilots—thirteen in all—took the stand to set the scene for Dameron's testimony.

Under meticulous questioning the forecasters explained why National Airport had been closed, leaving Dulles as the nearest alternate (on December 1, northeasterly cross-winds gusted laterally across National's north/south runway, but only quarteringly into Dulles' east/west runway 12), and, after agreeing on why a strip equipped only with a VOR DME navigational aid had been chosen over the main runway with its more sophisticated ILS (again, the wind direction), the controllers disagreed with their FAA bosses on whether or not they had received specific training for clearing planes on that approach: no, said the towed men; yes, said their supervisors.

Then two jetliner pilots who had landed at Dulles on the

morning of December 1 testified on the clearances they had received and their subsequent approaches.

American Airlines Captain Jan Minkler, heading for runway 12 half an hour before TWA Flight 514, said he was "surprised" at being cleared while thirty-odd miles out of Dulles. Flying the same course that Brock and his crew were to follow, Minkler said "we had no published altitude for that heading" and that he had looked at his approach plate and seen that the published routes over that area called for minimum safe altitudes of at least 3000 feet ("So we knew eighteen hundred was not the safe altitude to descend to") and had ordered his flight engineer to radio down for clarification. The transmissions between American Flight 300 and the Dulles tower were entered into evidence. There was a great nodding of heads around the hearing room.

"*Dulles Tower:* American three hundred, Dulles, yer radar contact cleared for VOR DME approach in runway twelve.

American 300: Cleared for VOR DME straight in to twelve—what's the last wind ya got?

Dulles: The wind right now, well, I'll tell ya, the wind's all over the place, let's see—it's outta the east and it's uh, swinging between zero nine zero degrees and one one zero degrees at two five [25 knots] and I've seen some gusts to thirty-five.

American: OK, thank ya . . . And, what's your minimum vectoring altitude out this far?

Dulles: The minimum vectoring altitude right there is four thousand feet.

American: OK. Could you give us a call when we clear Round Hill [the radio beacon east of the Blue Ridge marking the beginning of the final approach]?

Dulles: Yea. If you want, I can give you surveillance—you want surveillance to the runway?

American: That would be nice."

Jan Minkler had made a perfect approach and landed safely.

United Airlines Captain Robert J. Tyree succeeded Minkler on the stand and said that he had elected to come in nice and high that day because he had always been "cynical about approach clearances. All of us have seen errors or boo-boos made in clearances."

The FAA attorneys objected to his generalized appraisal of their performance. Admiral Thayer overruled them and asked Tyree to go on.

"I believe in checking [with ground controllers]," he added. "The pilot's responsibility is mine in more ways than one."

At 3:20 on the afternoon of January 30, 1975, Merle Dameron, blinking at the klieg lights—the television crews were back in force for his appearance—took the witness chair. Dressed for the occasion in a brown tweed jacket, plain brown lace-up shoes, beige shirt, and green knit tie, he somehow looked oddly meek and unprepossessing.

The NTSB's panel led him through his background: He was fifty-one years old, and had been a professional air traffic controller for twenty years—including, prior to his Dulles job, assignments in Alaska and Vermont. He had stopped smoking cigarettes two or three months before the crash, consequently gaining some weight, and, depending on his work schedule (he did not elaborate), he was in the habit of having one or two cocktails before dinner. The day before the accident he had played indoor tennis, spent the evening with friends, and had gone to bed around 11:30 P.M.

The questioning moved on to the controversy, previously testified to, over whether or not the Dulles controllers had ever received special training on the approach to runway 12. At issue was a statement Dameron signed, at the request of his union, the Professional Air Traffic Controllers' Organization, stating that he had received "no formal briefing" on how to land airplanes on that runway. PATCO, Dameron said, requested the statement "in case [the] FAA questioned anything I did." Now, he added, he would not sign such a letter because "it infers I was not prepared to accomplish or utilize this approach."

"At no time," he emphatically asserted, "did I feel I was not capable of understanding and interpreting the VOR DME approach." With that clear, Admiral Thayer recessed the hearing for the day. Outside the smoke-layered hearing room, night had fallen.

Back in the witness chair at ten o'clock the next morning, Dameron—soberly clad in a dark business suit, white shirt, and striped tie—faced questioning by the TWA contingent. Rapidly they led him up to the moment on December 1 when, after having cleared their jet for approach, he had looked at his radarscope and seen that the jet was at 2000 feet, twice as low as he had expected, but still flying.

"The read-out frightened me," Dameron admitted, "I'll say that." But, he insisted, he had operated "under the assumption that airline pilots ... the pilot—I shouldn't say airline pilots—the pilot is an intelligent individual—"

Interrupted by a brief, but quickly quelled flurry of laughs, Dameron continued, "—intelligent individual in my estimation, and the things that he does must have a reason. So, therefore, I saw something that concerned me greatly, but I had confidence that this man, if he was doing something ... he might be VFR for all I know, he might be flying visually with the terrain. In other words, he may have broken out [below the clouds]. I'm just saying all these things did not go through my mind right then, okay? But in retrospect, from my experience with pilots, that it was very possible that the man was flying under the clouds, with ground contact. So, until I can get confirmation of what is going on—by talking to the pilot—I wasn't ... you know, I hadn't flipped."

Still, he was asked, why did he wait until thirty-two seconds after the plane had vanished from his radarscope (and by then crashed) before radioing the crew to ask them for their altitude?

Dameron could not explain.

Nor could he explain why, during the five-minute and

thirteen-second period in which Flight 514 figured on his screen while sinking from seven to two thousand feet, he did not notice their premature descent.

Reluctant to let him off the hook, Dameron's interrogators tried another tack. How many times, the TWA panel inquired, did the controller's radar antenna pick up the bleep of Flight 514 during that period? Dameron tried to work out the mathematics: the revolving antenna scanned the sky west of Dulles fifteen times a minute . . . "Fifty-six," he said. Then, realizing that his computations were wrong, Dameron erupted at the pettiness of the question. "This is ridiculous!"

An FAA attorney rose to his defense. "In an *adversary* proceeding which," he emphasized, "this is, no witness should undergo . . ." Chairman Thayer gaveled the hearing to order, icily reminding the opposing sides that they would have ample time to get at each other's throats—in court. But TWA had made its point: had he looked at his scope in time, Dameron might have prevented the crash.

Thayer called a recess for lunch. Outside, it was raining. A few miles away, jets roared down over the Potomac to land at National Airport. Sitting in his taxi, a young cab driver watched them from the little park on the river's edge by the end of the runway.

"Some people," he said, "come out here hoping to see a crash. One day I guess they will . . . yup. I used to take my kid out here. He's four. But I just got a divorce."

He was not alone; another young man, in a blue sweater, desert boots, and khaki trousers, stood opposite the runway's centerline, and, mesmerized by the jets, leaned his fat belly into the wind.

Throughout his last afternoon of testimony Dameron, alternately under friendly questioning by his own people—the FAA—and less gentle prodding by the pilots' union and by counsel for the Aircraft Owners and Pilots Association (AOPA), a consumer group, rehashed what he meant by his clearance and how he had expected the crew of Flight 514 to follow it.

"I would assume," he said at one point, "that the aircraft would proceed on the three hundred degree radial to Round Hill [magnetic heading of 120 degrees—the straight-in approach to the runway]. Now, what altitude would maintain between the point that I gave him an approach clearance to the time he reached Round Hill, is the pilot's responsibility. He could have done several things in my estimation. He could have maintained seven thousand . . . he could have descended to any other altitude that he felt, you know, on the basis of information he had [on his approach plate, i.e., 3400 feet], and . . . I don't like to get into a chatty-catty type of thing, but an airline pilot, it has been my experience, is a very capable individual, and I know I am digressing, but this is the context that you have to realize that I [work in]—this is a capable individual. He has some very sophisticated equipment in his aircraft—I would say that Captain Brock, at the time I gave him his clearance, could pinpoint within five miles—and I'm guessing—but would be willing to bet that when he said 'thirty-four hundred' [feet was] indicated [on 'this dumb sheet' as the minimum safe approach altitude] that he knew generally where he was in relationship to Round Hill."

Dameron leaned forward toward the microphone. This was a crucial point. The pilot, Dameron was saying, knew that though he had been cleared for approach, the Appalachian mountains still rose between him and the end of the runway, knew that he had not yet reached the Round Hill radio beacon marking the safe, eastern side of the Blue Ridge. Data pried from the 727's wreckage supported Dameron's contention.

Had he considered Flight 514 to be a "radar arrival" or a "radar approach?"

Radar approach, said Dameron; a flight making an instrument approach but not one which had either asked for (as had American Flight 300) nor for which FAA rules required controller monitoring all the way to touchdown.

Were the TWA jet to have been classified in his mind as a "radar arrival," Dameron said he would have, naturally, given it an altitude restriction until it had crossed Round Hill. But since the crew was in charge of its own navigation, "the pilot has the responsibility for [his own] terrain clearance."

As with the term "cleared for approach," it became obvious that there was a grave potential for misunderstanding between controllers and pilots as to when a flight was or was not receiving the maximum radar surveillance afforded by a radar arrival.

Again Dameron was asked why he did not spot Flight 514's premature descent. Again he insisted that there was no reason for him to be monitoring that flight but that, to try to answer the question, all he could say was "in discussing this afterward, and believe me, this is a horrible thing, and I've discussed it with a million controllers, I was amazed that I did not observe this aircraft's altitude."

The hearing room was hushed. For the first time, someone had admitted that there was more to all of this than technical jargon and acrimonious legal double talk; someone had actually spoken of the "horrible thing" itself, the reason for this hearing: the deaths of ninety-two people—thirty-eight women, fifty men, and four children.

Dameron swallowed. A small silence ensued. It was obvious to everyone in the room that, wrenching as it had been, Dameron had spoken the truth; he simply had not noticed. And that that was his only, agonized, answer.

At 4:05 P.M., Thayer excused Dameron as a witness, then, apologizing, called him back.

"I beg your pardon, Mr. Dameron. I will ask you the question I have asked everyone else. And that is, if you have anything you would like to add at this time?"

Dameron sat down. "It has been my observation that there is, seems to be some confusion between what the pilot thinks and what the controller thinks. In this case, in my clearance, I thought it was very clear and concise. I did not feel the clearance was confusing. I thought I knew what the pilot would do. I felt, in turn, that he felt the same way because if he was confused about the clearance, I feel that the man would have asked me.

"Now, we have heard a lot of things during these proceedings about Jeppesen charts, NOS charts, TWA flight manuals . . . it goes on and on and on. These papers or publications essentially present all the same information, but they present it differently. And I—it is an observation [and] I'm not sure it is appropriate—but it would seem that if we could standardize our publications so that the traffic controller is using the same book to control [the pilot], so that everybody is using the same publication, you don't have fifty different things that you are referring to, then people—both using the same reference—they know what is going on.

"And that," he concluded, "is all I have to say."

Dameron stepped down from the witness chair and walked out of the hearing room. Obviously shaken, he could only smile wanly at the backslapping congratulations on his testimony offered by other controllers who were, no doubt, profoundly grateful not to have been in his shoes.

With Dameron gone, the number of television crews and the general audience dwindled. But there were some who hung on with tape recorders or just searching faces. They were the relatives and friends of the dead and they came to see if, for once, death could be explained, neatly dissected, and laid out along with the graphs, maps, and charts on the exhibit tables. Disillusioned, some left after a few sessions as did Alice Noxon who remembered "sitting there, and they are going back and forth and you hear everybody's little side— and I walked out saying 'the crash couldn't possibly have happened' . . ." She had lost both her mother and father. ". . . The pilot was exactly right, everything about the plane was exactly right, the controller did exactly what he was supposed to do, the weather was a little iffy but it wasn't that bad . . . and I'm sitting there going, 'Well, the mountain jumped up and bit the plane.' "

And some, like Alan Clammer, Tom Safranek's friend, were there, along with the flight engineer's father and brother "because I didn't want to see TWA wiggle out of their responsibilities, I didn't want any whitewash. Knowing TWA as fully as I do, I fully expected there would be some kind of whitewash. And they made some attempts, as you saw at the hearings. I mean everybody was running for the hills, nobody was willing to accept the slightest amount of responsibility.

"Our director of flying briefed one of our captains in his hotel room and told him that at no time in his testimony would he refer to the radar altimeter as the radar altimeter— which it's called in every book we have—but would refer to it as the 'terrain avoidance' altimeter, thereby injecting TWA's phony philosophy that they have adequately equipped and prepared their airplanes and crews with proper instrumentation to avoid terrain, which is certainly not the case. For example, the NTSB has been urging the FAA for years to make ground proximity warning radar devices mandatory. TWA's major overseas competitor, Pan Am, saw the light a year prior to this crash, when they lost four airplanes in one year. This is not excusing the pilots who flew into the sea or into the ground . . . but whether it was their fault or not, the ground proximity warning radar device gives you a last chance—it gives your passengers a last chance. If, for any reason, you're too close to the ground, all you got to do is pull the nose of the airplane up and people live instead of die. It's as simple as that."

Shortly after Flight 514's crash, the FAA put out a mandatory note, ordering all U.S. airlines to have ground proximity warning devices installed in all their planes by December 1, 1975. A TWA spokesman said it would cost them $2.5 million—5.2 per cent of the $47.78 million the airline admitted spending on publicity and advertising in 1974.

And the refrain went up, once again. "Why does it take a major disaster to get the FAA to enforce something everyone knows ought to have been mandatory ages ago?" Unfortunately there isn't any simple next verse.

Clammer: "My initial ambition [in going to the hearings] was that I was going to write a book on airline safety ... analyze it differently than any other book which had been written because I've got them all, over there in the bookcase, and there isn't one of them, in my opinion, that's worth a damn. And I was going to make the book center around Flight five fourteen and the causes of the accident because I was quite certain that those responsible would really come off pretty clean, owing to the NTSB's previous record. I must say I was surprisingly impressed with their performance at the hearing, to some extent.

"I also had many personal issues regarding the airline's safety that I think are dangerous, things that are going on every day."

For example: "The radar altimeter on that airplane [a Boeing 727] and on all of our airplanes, except our L-1011s, only has a five-hundred-foot readout ... unlike every other carrier in the goddamn world. The transmitter in the belly of the airplane has a twenty-five-hundred-foot capability, but the gauge itself only goes to five hundred feet. Cheaper, that's right. We don't have strobe lights in our airplanes, like everybody else does. We've just got that little red, rotating beacon which is nothing. Everybody else's got strobe lights. We're so busy painting our airplanes new colors and all that bullshit that we can't afford strobe lights. We've been asking for them for years ... These are things the public should be aware of. I have no fears at all for my position at TWA. I couldn't care less if they bend me out of shape."

During the hearings, Clammer went up to the crash site. "There were some tears ... but there's been lots of those anyway." He isn't the kind of man one easily imagines crying. Tall, blond, and trim, he's both precise and gentle in the way men unburdened by insecurities can be. "It had been long enough after the crash that the shock had worn off. I knew he [Tom Safranek] never felt any pain. The lights went out, that's all.

"I wasn't looking for any facts; I just wanted to be

there . . . It was cold and dreary, but no snow . . . They had raked a lot of stuff right into the ground—they didn't clear everything away. I left a couple of daffodils."

Without daffodils, I went up there on a day much like December first. The fog curled around the brutalized trees, icicles dripped from the rock ledge which gutted the plane. The basalt was scarred, topped with snow. Someone had thrown up a wire fence to keep out souvenir hunters, but my dog, a young black Labrador, burrowed under it and sliced off through the wet snow. She hesitated, for a moment confused, but then, nose down, followed some ghastly, buried trail. She stopped at the base of one of the trees and dug furiously. I called her back.

Standing on Route 601, looking west down through the broken boughs, it somehow was easy to imagine the airplane plowing on through, in slow motion, silently . . . the nose dipping through the trees, the wings, blunted now, the green-tinted cockpit glass, shattered, looming larger. But the impact itself was impossible to reconstruct—the explosion of flesh, fuel, and metal. This is no place for an airplane, this nondescript little portion of the Appalachian Trail.

On February twenty-second, the day after the close of the NTSB hearings, I was offered a ride over the crash site and took along a notebook:

Take-off from College Park, single-engined Piper Cherokee. Climb to 1500 feet, cross the Potomac at River Road, heading northwest. Leave Dulles to the left. Come up on the Blue Ridge slightly south of Route 50 at Snicker's Gap. The ridge line, beige winter trees, khaki ground. Make three passes over the crash site after having a tough time finding it . . . a small clearing in the trees. Head west over the Shenandoah to make the same approach as TWA 514. VOR DME on Radial 300. Dulles tower says "Okay." Bank left. Fine crisp morning. Dulles asks for a few course corrections. "Okay, you're on it." Ridge line ahead and below. Glimpse of gray river, again. The hills slope up gradually. Coming in at 2700 feet. Angrily Dulles says, "At 2700, you are below my minimum vectoring altitude for that area." It's 1:07 p.m. If the Dulles controller had said that to TWA 514, I wouldn't be here. Coming in over the foothills. Updrafts. The crash site is ahead, now directly below. Insignificant. Have the feeling that if the trees hadn't been there, the jet would have made it . . . The hills level off at the ridge line, a gentle rounded hump. The swath is small, looks as though some careless lumberjacks had been at work. The path through the trees brings it all into proportion, a small clearing in the forest, a little void in both the landscape and in the consciousness of those left behind. From up here, the hill looks soft, furry. It isn't.

Over the ridge . . . long flatland, patchwork of winter fields, small lakes, animal drinking ponds. Completing the approach: can't see Dulles for awhile, then can—a wide band of white, the runway. A big black "12" on the concrete. Landing is like going uphill. [Flight] 514 never got this far.

In the air again, before mid-runway. Back toward College Park. The spire of the National Cathedral pokes up through the sun glint. Flying an airplane, despite the technology, the navigational aids, the computer, is still keeping the bloody thing in the air, away from the ground. The old sky diving motto comes back: "We are trespassers in the sky." . . . Land, narrowly missing yellow oil drum.

Down at the gleaming NTSB headquarters opposite the Hirschhorn Museum the experts sifted the data, apportioned the blame. In a year or so an "accident report" would be issued, a tidy, blue-bound booklet, explaining it all.

CHARLES ("DAVE") PIERCE, 23, BODY BAG #129

After he had satisfactorily centered and pinned his lieutenant's bars on his freshly laundered uniform shirt, dressed, stacked his extra change on his youngest brother's nightstand, and shot a last look around his room with the carved wooden scouting eagles on the wall, Dave carried his gear to the car, stowed it, and walked back to the house to kiss his mother goodbye. He first refused, then sheepishly accepted the twenty-dollar bill she proffered.

Bettye Pierce remembers the exchange well; it was the last time she saw her son. She also remembers, though less clearly, the rest of that afternoon, filled by fruitless telephone calls to TWA offices in Indianapolis, Cincinnati, and Chicago, the fighting back of tears at the busy signals, the blind hope that the "C. Pierce" identified on newscasts was someone else. "C. Pierce" certainly did not sound like the name of a young marine lieutenant, her son. An afternoon of waiting for a phone call from Columbus (where, because he flew standby, Dave might have been bumped from the flight), from the Quantico barracks in Virginia where he was headed, from anywhere. It was TWA that finally called, at nine that evening, politely asking for clues which might help identify the body. It was already bodies, not people.

"He took such good care of himself. He believed in body beautiful," Mrs. Pierce might have told them, "and then somebody made a mistake. *Somebody* made a *stupid* mistake!"

She didn't though; she was helpful instead. No scars, but he'd been in uniform, dogtags and all.

Clues . . . the Pierces themselves. Bettye, the mother. Clear blue eyes behind glasses, curling brown hair, solid legs —the kind of woman who can stay on her feet all day long and not complain of a backache. Who works for Sears Roebuck in the computer department. Who is deeply resentful, though she tries terribly hard not to show it, that someone goofed on their job and killed the boy for whom she'd gone back to work to support through college. Harve Pierce, her husband, a mild man who works for the phone company and likes to tinker with things like CB radios; who keeps the same car for years and years, unashamed, unfettered with the keeping-up-with-the-Joneses syndrome. And this in a car-happy town, living in the car-happiest of neighborhoods, Speedway, two blocks from the Indianapolis 500 Big Oval ("Dullsville, I believe it is called after May of each year," Mrs. Pierce was to write. "Personally I like ruts. The deeper the better— ha!!"), row after row of white single-story houses sprouting television antennas, peopled by what Bettye laughingly refers to as "WOCs"—White Occidental Christians—"rednecks," pause, smile, "like us." The apocryphal neighborhood-you'd- like-your-kids-to-grow-up-in; schools and playgrounds within walking distance, softball on summer nights, enough churches (Baptist, Methodist, Presbyterian), pizza parlors, and speed shops to be a Middle-American nirvana to Jesus-crazed, pizza-starved car freaks. The kind of neighborhood where Ted, Tom, Marsha, and Dave Pierce grew up, none of whom, however, fit the town stereotype.

Ted, the youngest, thirteen when his brother was killed, a budding swimming champion (chin-up bar in the doorway

to his room) and entrepreneur who built a plank bridge over
a muddy ditch near the Oval and gratefully accepted "contributions" for passage on race day. Who nailed up his own,
newly acquired, wooden eagle-scout eagle next to Dave's.

Marsha, just married, relieved she says, that her husband
and Dave finally liked each other on that last Thanksgiving
weekend, blushing just a little when she says that Dave had
made going away to college easier. "Because I had an older
brother, I knew what everybody was talking about. It seemed
like he always called at the right time."

And Tom, whose life was disrupted and thrown off balance by the crash more perhaps than anyone else's in his
family. A handsome young man of twenty, with his mother's
brown curly hair, shy, struggling to explain: "Well, it was
something that I hadn't realized, that I didn't know when
he was living, you know, a lot of people have heroes, like
in athletics and stuff... I didn't realize I was patterning
my whole life after him. And so afterward I really thought
back and gosh, your model's not around anymore.

"I just don't know what I'm going to do anymore. OK,
just to give examples ... like he was in scouts. I started in cub
scouts and he made eagle scout and that was just something
for me to make and when I finally got eagle scout, he went on
to college and he was interested in police work, and I was.
And then he was getting really started in the marine corps,
officers training, and I had everything ready to go in, just
like he had. But then, after the crash, I just kind of forgot
about it. I had everything signed and ready ...

"He said, 'Don't ever listen to anybody else, trying to tell
you to do something. Only do what you want to do.' We
talked one night, just the two of us when he was home.
'Cause I was kinda apprehensive about going into the marines
and stuff, I was kind of scared and stuff and he was just
saying all this stuff, just trying to calm me down. It was
really relaxing to talk with him. See, he had a sense of humor.
He could tell you something trying to be serious and you'd
just break out laughing at the way he'd say it. He knew

about my stomach problems and he even told me stories so that I just died laughing off things that had happened to him. Like he was always one for physical fitness, jogging and stuff, and he was out jogging once and he thought he just had gas pains so he decided to let it out; but it was more than gas and the last mile was the hardest he'd ever run.

"So it was just stuff like that . . . I could never accept that everybody, you know, had the same problems that I did."

Clues . . . Finding himself more a creature of the fifties than of the long-haired activist generation to which he belonged, Dave Pierce had dreamt of becoming a marine the way other boys long to be airline pilots or firemen. Conscientiously completing his ROTC courses at Vincennes University, he was graduated in the spring of 1974 with a degree in criminology and forensic sciences and an about-to-be-confirmed commission as a second lieutenant in the United States Marine Corps.

And, in that spring when the very existence of the corps was hotly being contested in Washington, with Vietnam crumbling and American involvement in Southeast Asia a nightmare best forgotten, Dave Pierce drove down Cord Street in Speedway, Indiana, wrapped in a bedsheet so as not to dirty his new dress whites. Back from the commissioning ceremony a few hours later his whites were wrinkled, but two slim gold bars gleamed on his shoulders. Proudly, he left for Quantico the day after The Race—Johnny Rutherford won that year—to report for basic training.

When Dave came home for Thanksgiving six months later, fit, tan, and a week short of finishing boot camp, Bettye Pierce found that her son had "come of age." The "big talk and big front" as she called it, the rough talk and swagger he had affected in college ("it was like he was practicing to be a marine all the time") had faded, she said, replaced by the correct, if slightly stilted, code of conduct instilled by the corps.

She had found him to be more "broadminded" than she

remembered; yet he was also certain that "he could have settled the Vietnam thing in no time at all," sorry that he had not had a chance to do so, but optimistic that there would always "be something" happening in the world which would require marine intervention.

Bettye laughs indulgently as she recalls her son's martial zeal, finding no paradox between his so-called broadmindedness and his full-fledged bomb-the-bastards-back-to-the-stone-age attitude. After all, this is the state of Indiana, "Hoosier" country—that peculiar nickname which, myth has it, originated in pioneer times when a strange knock on a cabin door would elicit a gruff question from within of, "Who's-yar" brother, mother, or whatever kin struck the fancy of the questioner. This is the state where the Ku Klux Klan once flourished, where in 1958 Robert Welch founded his John Birch Society, where state fairs and 4-H clubs remain sacrosanct, where the fall of Richard Nixon is still widely regarded as the product of a demonic eastern left-wing conspiracy fomented by *The Washington Post*. And where ethnic jokes are told, not about Italians, Poles, or Mexicans, but about the folk from Kentucky across the state line—"Question: Why couldn't Jesus have been born in Kentucky? Answer: Because they couldn't find a single virgin or three wise men in the whole state."

Yet, whatever their political persuasion or their attitude toward welfare, Medicaid, unemployment compensation, or the future of New York City, people like the Pierces are the flesh and blood, the meat and potatoes of the broad belt of real estate between the Ohio Valley and the Rockies commonly thought of as Middle America. Raised during the Depression, largely self-educated, hard-working, and frugal, protective of their accomplishments, they struggled to understand their children's fascination with acid rock or, as it were, the hallucinogenic drug itself. And when a young man emerged apparently unscathed from four years of college and was willing, indeed anxious, to chop off his hair and put on a uniform, they were proud.

These are people who believe in the System, and when it fails, when as basic a link in it as a common airplane ride fails, they are shocked. "Somebody made a stupid mistake," says Bettye Pierce. Blankly, they turn to the System for answers and find that there are none; even if they obtained a copy of the National Transportation Safety Board's Accident Report (when it finally emerged), it is more of an alibi than an explanation. So, they seek the comfort of their churches, some, hesitatingly, the help of psychiatrists. But there is no rationale for getting killed and they learn this. There is no reason, no saving grace—just the fact that it happened. Eventually they—all of them without exception—turn to Associated Aviation Underwriters, the keepers of the treasure. In carefully computed dollars and cents they haggle and are, eventually, paid for their loss. Though the money effects a truce, it does not absolve anyone (the airline and AAU bluntly refer to the settlements as "buying our peace"), and obviously cannot replace the son, daughter, mother, or father mangled on the Blue Ridge. Nor is it designed to. It simply represents the System's imperfect way of balancing accounts.

There was no serious question in the Pierce family on whether to settle or sue. The insurance adjusters from AAU were "as nice as they could be," according to Bettye, who was particularly touched that they returned Dave's wallet and his marksman's medal. The wallet was intact and the medal was slightly charred, reminding her of the violence which had come to pass on Weather Mountain.

AAU first offered the Pierces $25,000, which they refused. Tom, aghast at the offer, had turned to his parents when he heard about it and blurted out, "Yeah, twenty-five thousand dollars . . . Is that with or without a free jar of vaseline?" They had hushed him, embarrassed by his youthful choice of metaphors. By July, seven months after the crash, they had upped it to an acceptable $50,000. Still, the elder Pierces are loathe to discuss the settlement. "It's like putting a price tag on him," says Bettye.

A year after the accident, the Pierces signed the deed to the land they bought with the settlement money. Thirty-six acres and a farm. Tom and his sister contributed their share of the settlement to that end. Ted, legally unable to follow their example, had his money put in trust toward his education.

Tom, although relieved of the burden of spending money about which he has qualms, saddled himself with guilt, using his brother's death as a ledger for his own emotional accounting. After the crash he left college, moved out of the family house, and went to live in draughty loneliness with Tina, his St. Bernard puppy. He says he then wished he had been on the plane instead of Dave, "who had everything going for him —I just wanted so hard to change positions . . . I consider myself a total waste." Eventually convinced by a psychiatrist "that I couldn't have stopped the plane from crashing, that I couldn't run away from problems," and that feeling guilty for living a normal life ("you know, going to the movies, things like that") was unrealistic and self-destructive, Tom recognized that "I just had to accept what had happened." He still lives alone with Tina, still has his brother's picture in his living room, and still (carefully) wears Dave's marine fatigues, but he has returned to school and has even bought a Jeep.

While he struggles to accept Dave's disappearance, so do his father and mother. A year after the crash, with the farm bought, the settlement settled and the first anniversary passed and survived, Bettye, collecting her thoughts and memories, wrote me:

I really wonder about us. Are the Pierces of the complacent society . . . the silent majority . . . the Do Nothing Group? Whenever I ask myself this I always have the same answer. You can only kick a dead horse for so long before you know it isn't going to move. When a finger is pointed, it goes full-circle of "Who me????"
You only have to look so far to get a very good idea of

what I'm saying. . . . Indiana is represented by one "hayseed," Sen. Hartke and one "pretty boy," Sen. Bayh. If they wanted to do something [about aviation safety] they could do it. However, it doesn't seem to fall into the scheme of things. Newsletters still come here addressed to David from Sen. Hartke's office . . . Isn't that a kick in the kimono???? Someday when the day is dark and gloomy and my spirits match the day, I'll let fly—POW!!"

Gee, I didn't intend to get political.

Tom is back to the grind—full time at college. He seems contented. He said the other day that he didn't realize how much he had learned in such a short time and he hoped he could retain a part of it at least. He is sticking with criminology with hopes of becoming involved with keeping the laws some way.

Ted is wrapping up his swim schedule with just a couple of meets to go. He hopes he will have accumulated enough points for a reserve letter.

His long-range plans include a canoe trip to Canada in July with the Scouts, and Drivers Ed. On the latter, I can hardly wait. One more child to tell me what a lousy driver I am . . . They never seemed to mind when I'm chauffering them all around God's Green Acres.

Harve just returned from Savannah, Georgia. He was attending a three-day seminar for Indiana Bell. This was the first time any of us had flown since the accident and I tried not to overreact and remain calm and tell myself the good Lord wouldn't let tragedy hit twice in the same way. My stomach was in knots until he called saying he was OK. "Oh ye of little faith . . ."

SHEILA REGAN, 33, BODY BAG #39

The men wander in out of the desert dusk pulling on their jackets because of the air conditioning. They order beers or martinis and talk of advertisements bought or canceled, about the ex-croupier with cataracts now selling mobile homes in Barstow, about the water table and the tourist trade, about Petula's comeback and, for the edification of an outsider, about their city.

"Half this country still believes Las Vegas is a strip downtown and nothing else. They still don't believe that real people live here, people who go to work, to church, who bowl, who have children, just like anyplace else."

". . . *not* being in Vegas is when you miss it most. Not that you do it, but there's nothing you can do here at three P.M. that you can't do at three A.M."

From the end of the bar, in a Highland brogue two decades out of Edinburgh comes the observation that "You have to learn to live with temptation here, particularly as a single person. If you can't learn to live with that in six months, then you'd better leave."

Sheila Regan drove into Las Vegas in the spring of 1965, following a man. She was twenty-four and in the middle of her first serious and illicit (the fellow was married) affair. It did not last. He moved on, she stayed.

Sheila, they say at the bar, ah . . . They all knew her, these ad-men, flacks, promoters, and general hangers-on who drink at the Las Vegas press club, and some of them knew her well; it shows in their eyes.

The Highlander again: "She wasn't one of those girls who came up looking to make a quick buck, eh Judy?"

Judy Fensterman shakes her head; they had been a team, Sheila and Judy—the heavy-boned Irish girl and the sunwizened Vegas sharpie.

"Pretty girl though," offers a martini drinker, "tall . . . attract attention when she walked in a room."

Judy nods confirmation.

"And let's face it," the Scot says, "you don't expect it for a woman to go into the FBI. My goodness."

"Sheila, in the *FBI*?" This from one of the few ladies listening in.

"Sure . . . didn't you know?"

"One of the first women ever appointed," chips in the bartender. "Right, Judy?"

"The third—*ever*."

General wonderment. Drinks all around.

Growing up Catholic around North Hollywood—parochial school and a four-year scholarship to Immaculate Heart College—Sheila studied mathematics and physics (movies, despite the obvious dreams, were to be seen, not acted in), supporting herself by secretarial work at the *Valley Times*, a small weekly with offices on Magnolia Boulevard. At twenty-one, she enlisted in the marines and wrote her brother Jim, then serving with the corps in Hawaii, that "I decided I'd take off and go to Quantico for the summer." Four months later a thinner, wiser Miss Regan emerged from Officers Candidate School, declined her commission, and returned to California for a last year of college. Graduated in 1963, she went to work full time for the *Valley Times*, a job she held until she came to Las Vegas.

Leaving only the sad ones who never eat dinner, the crowd at the press club thins. Time to go. Judy insists that we "do" Sheila's Las Vegas. The tour begins at a Mexican restaurant, the Macayo. Margueritas, Judy says, Sheila was big on "Margs." We order Margueritas.

". . . four-two-five. We had the perfect formula. Four parts Tequila, two parts Triple Sec, five parts Sour Mix. Of course, after the first and second you can't taste it anyway . . . When I get to where Sheila is, she'll have a Marguerita waiting, she'll have it all organized when I get there." Judy has a huge, white smile, speaks in a permanently breathless husky voice, has a son who made the All-State basketball team, and is, roughly, in her forties. Ferociously protective of Sheila's memory, she polishes the joyous anecdotes and omits the nights of mutual commiseration following her own divorce and the bitter, rasping end of Sheila's first affair. "Basically a good percentage of the people that you know are married. Okay. Now there were a lot of married men that, you know, given the chance, they'd have thrown everything away. Well, you don't let them do that, or get to that point. There are a lot of men who idolized her."

Within three years of her arrival in Las Vegas, Sheila was working for the Hughes Tool Company. Starting out as a receptionist, she quickly moved up to become secretary to Francis Fox, who ran Hughes's airport and airline operations. Partly because she had kept the books and files—she knew where the money was and how it had been spent—and partly because she was hard-working and discreet, Sheila survived the 1970 coup within the Hughes empire in which Robert Maheu, Fox's boss and the man in charge of the entire Nevada complex, was unceremoniously ousted. The incoming clique took over and promptly hired International Intelligence, Inc. (Intertel) to clean house and put the fear of God and Howard Hughes into the remaining employees.

Sheila's brother, Jim, a detective sergeant for the Santa Barbara sheriff's office, described Intertel as a private investigatory agency which "specialized in keeping organized crime out of large corporations. It was started by this guy who used to be a prosecutor for the Justice Department and it's got a lot of heavy people. In fact, the guy that started it is the one who discovered—remember when that phony Hughes autobiography came out [Clifford Irving's]?—Peloquin is the one that found out where all the money went over in Switzerland."

Robert Peloquin, Intertel's owner, founder, and president described his company as "a security consultant firm." "Our average client," he said, "would be some corporation who might have a large inventory loss and who asks us to come in and set up a system to prevent it from happening again.

"We were hired in early December of 1970 to provide

management direction and services for Hughes's Nevada operation—which was mostly hotels and casinos—because they
were changing management," Peloquin explained. "Sheila
had the choice of either staying in Nevada or going to Los
Angeles and she decided to stay and continued working for
the Tool Company for I guess five or six months under our
direction. We then opened an Intertel office in Las Vegas
and picked her and two other ladies to shift over and work
directly for us. She was kind of a joint secretary/administra-
tive assistant. She did what she was . . . Well, she did what she
was told to do. I'd say secretarial work was about fifty per cent
of her function, the other fifty per cent she kept the books in
the office and devised a central filing system."

With the Maheu firing, attendant dismissals, lawsuits, and
federal and local investigations, keeping the books for Hughes
and its rent-a-cop guardian was a sensitive position. "Sheila
was," Peloquin said, "very discreet. Part of the job." Though
she did not get a chance to do much investigating, he
said, "except if someone needed a girl to accompany him to
make it look natural," Peloquin sensed that "she absolutely
loved her work, no matter how much on the fringe she was
of it. She loved being in Intertel because of the type of work,
the type of people"—the policemen, private gumshoes, and
various agents of Federal agencies who are to Las Vegas a
little what stockbrokers are to Wall Street, ski instructors to
Aspen, or heart specialists to Miami Beach: the hard-core
service population that struggles to keep paying clients more
or less honest, rich, happy, or just alive.

"The office in Nevada was a fairly social office; they'd
have a lot of parties. Sheila did hang around with older people, a rather broad spectrum. She had political friends, people
at the city clerk level, and she had impressive friends at a
relatively higher level. An example—the state senator from
Las Vegas, a guy by the name of Chick Hecht, she was
friendly with him. He had a group or a club or whatever you
want to call it, called the ex-CIC that was made up of people

who had been in the army counterintelligence at one time or another, and she was kind of their honorary secretary ... She just liked that ..." Peloquin laughed, "... liked the law-enforcement type."

Attempting to explain Sheila's choice of companions, Peloquin explained that "Las Vegas is a very small town when you get down into it, a very close town; people kind of stick together because they're all on the defensive, psychologically. Everybody is dependent on gambling and you get a slight smell of sin in there, so everybody slightly defends it and they think people are looking down on them because they're in that industry. So even people like the FBI or the IRS agents—they got a big office; Christ, there's about one IRS agent for every two people—are slightly on the defensive in that town."

Two more Margueritas and the check.

Judy Fensterman recalls Sheila's brief political career which began with parties at the Las Vegas Young Republicans Club and grew to a 1970 campaign for a seat in the Nevada State Assembly. Judy had acted as her campaign manager. "We had no funds, so we sent to California to Governor Reagan's headquarters and got his bumper stickers which said R.E.A.G.A.N. So we cut out the 'A' and had them all over our cars."

Peloquin: "One guy with whom she was very friendly was Jack Jones, who was basically, I guess, her political advisor— he ran a real estate agency and dabbled in local politics—who would map out campaigns for her. Now at the same time, she'd try and angle him into getting maybe an advertising thing from Hughes or what have you. Every time I'd turn around, Jones would be under my feet. Finally I caught on, and I gently told her, you know ... to lay off."

John Jones clearly remembered Sheila Regan. He handled several local candidates' campaigns in 1970, but "we did it for Sheila as friends, we didn't charge her. She came into my office one day and I told her, 'Sheila, you got to have some positions, a platform, to make it look like you've thought it out, like you know where you stand.' "

Good idea, said Sheila. Jones rolled some paper into his typewriter and tapped away, and Sheila instantly emerged, according to Judy (because Jones said he could not honestly remember what the hell he'd written), with a stance opposed to "so many attorneys in the legislature," and against abortion.

"Certainly no big women's lib thing," said Jones. "But I told her she'd lose and she did. Now, the second time around, in 'seventy-two, she was a lead-pipe cinch to win. They had redistricted, her district was smaller and she ran unopposed, there was no Republican primary and her opponent, the Democrat, was, well ... let's just say, I didn't see any reason why she couldn't beat him. Then the FBI thing came up and she had to drop out of the race."

Peloquin: "When we closed down our Las Vegas office, Sheila and the two other ladies had a choice of either looking for another job or shifting over to Los Angeles. At that time Tom Carney, the head of the Nevada office, had heard that the FBI was recruiting or was discussing recruiting women. Sheila still didn't want to go to Los Angeles and she was overeducated for the particular job she had been fulfilling, or the jobs she had been fulfilling. She really had a lot more talent. As with so many women, I guess you kind of slot them into the secretarial mold, and Sheila didn't want to be slotted into that for the rest of her life, and I'm sure was bugging Carney to help her break out of it. Pat Gray was then head of the FBI and he made a speech or something, and it occurred to Carney that when this FBI thing came along that it might work out.

"Sheila was a very square person, if you want to put it that way. The minute I heard that Carney was trying to push her into the FBI, the thought occurred to me that there's an ideal place for her. If there was any girl I knew who would do well in that, it would be her. Just by her nature she was a person that follows proper orders."

Judy insists that the next step on the Sheila Regan Las Vegas tour must be the Silver Dollar. Fire-glow in the night sky. Judy squeezes her bright red Toyota through the crush of cabs nudging toward Caesar's Palace. On the sidewalks, old women on special off-season discount junkets grasp plastic cups of nickels and dimes, while hookers cruise. Judy swerves off the Strip. Blocks of darkened houses surrounded by stubby lawns and the cooling asphalt of playgrounds bear witness to the "other," what she calls the "real," Vegas.

"Mr. Carney heard that they were accepting female candidates for the Bureau. In fact, we hadn't heard of it, I mean, just as citizens I had never heard of it, but a lot of the men at Intertel had worked for the Justice Department, a lot of them were former agents, they had the pulse of the thing. So he called down for her, found out what the requirements were and then she went down to get the forms—which are exceedingly endless—and she filled them out and sent them in. She had to get a physical, which she took at Nellis [Air Force Base]."

Six weeks after running her background check, a local Las Vegas FBI agent called Sheila to tell her, unofficially, that she had made it. A letter would follow. For days, if she had to leave the house before the morning mail had been delivered, Sheila would tape a note with a dime and a telephone number to her door asking the mailman to CALL! The letter came on a Saturday. "We sat on the floor and read it," Judy remembers. "First of all it says 'Congratulations,' and then—it was rather lengthy—it said she had to report to Quantico, I think in about two and a half weeks, and said like you have to bring so much money, you won't get paid for a certain amount of time and . . . the funny part was the two jock straps—she thought she'd have to wear one on each arm—and then she had to have shorts and a tee shirt and a sweat shirt and tennis shoes and all that garbage."

Though disappointed that she had not been named to the first class of women agents ("the one with the former nun and the former marine," says Judy), Sheila worked assiduously at getting into some semblance of physical shape.

"Sit ups, push ups . . . She was tough, I mean she could do these push ups. I think those are impossible, to lift your rear end . . . and running—you have to run two miles in fifteen minutes or something like that, and if her feet hurt she'd say, 'Well, yours would too if you carried all this [weight] around.'

"Sheila was a voracious eater. She loved steaks, medium rare, and she always cleaned her plate. Always. She'd do everything but lick it. She did drink the dressing out of her salad. And double chocolate sundaes. But, sometimes, you'd go through periods—we'd have dinner parties for two [in Las Vegas]. We'd have candles and put out all the best china. One time I was over at her place and this guy asked her out and she left. I made dinner by myself. That was the beautiful part of our relationship: if you got a better offer, the other would be understanding."

So Sheila finally went off to join the FBI. Towing a U-Haul trailer behind her aging, metallic-green Pontiac, Sheila and her dog, Brandy, chugged cross-country from Las Vegas to Washington through the heat. On September 11, 1972, with Brandy safely boarded in a kennel, Sheila headed south on the familiar stretch of Interstate 95 to Quantico, the sprawling marine base where the FBI had its training academy. Ten years . . . Sheila must have inwardly smiled at how far she had come since she had last driven down this highway—all the jobs, California, Las Vegas, Hughes, Intertel, the men, the death of her father, the crazy weekends with Judy, the Margueritas by the pool . . .

She drove past the marine tank farm, fuller now than in 1962, the rifle ranges shimmering in the heat, the white-washed barracks, the skinny boys in fatigues looking up at the girl in the green car with the rotten muffler. It was going to be boot camp all over again—calisthenics, stern instructors, manuals to be memorized, drills—but when it was over she would be FBI Special Agent Sheila J. Regan and not some lonely, exhausted girl anxious to get back to the familiar classrooms of Immaculate Heart College in North Hollywood, California.

"She made a lot of good friends at the Academy," Judy says. "She always wrote and told me that she had guys picked out for me. Fantastic, you know, just made in heaven. 'Eagles.' See we had . . . certain men were eagles; the only way you could describe them was that they were outstanding, they were eagles. Also, there's a lot of foreign students from other governments and there are FBI short training classes for police departments, lots of people beside your own class . . ."

It is Thursday and the Silver Dollar is crowded, hot, and deafening. Country Western and hard rock. It is not a touristy place; some of the men look like mechanics and truck drivers—cigarette packs tucked into the rolled-up sleeves of tee shirts—and others look like sales clerks trying to look like managers, with their double-knit blazers in red, green, and yellow patterns. The women mostly seem tired and lonely. Judy wants to dance and immediately kicks off her shoes. Wishful wildness. We dance, finally sit down, gulp cold beer. The Macayo food refuses to lie quiet.

"Sheila set up a bar in her room at the Academy and liquor wasn't allowed. She asked me to send her stir-sticks and napkins, and then they did a search or whatever they do and she was uncovered; she had to remove her bar from the premises."

Graduated from the Academy on December 19, 1972, Sheila reported for duty in the Philadelphia field office after spending Christmas with Intertel friends in New York, relatives of a United Airlines pilot whom, Judy says, Sheila dated from time to time. "She called me on Christmas Eve. I was so proud of her. Really. I thought it was a big deal her being an agent, but the worst part of it was I couldn't talk to her a lot about it."

The FBI was equally stingy with information, even though any case on which Sheila worked has now long since been closed, with innocents freed and the guilty convicted, and perhaps the reverse as well.

Peloquin: "She had one case she was all excited about where they used her as an undercover . . . She was supposedly a gun moll someplace in Pennsylvania. She was so excited that it had worked . . . It was a group that went around committing high-class robberies, where they hit places like the Main Line of Philadelphia, and there were also murders at the same time, as they went into the houses. I don't know if I got the story straight, but there was some guy that was coming out of Florida or something, coming up to Philadelphia, and she was supposed to be the contact to get him a car, and she was supposed to be the friend of some other guy's broad, and they had her all wired [microphones in her brassiere, says Judy], and it apparently worked very well. Yeah. She was all charged up about that."

Jim Regan: "She got a flat tire somewhere in Philadelphia and it was right in front of this warehouse and this guy comes out and offers to fix it and drags her car into the warehouse. He's just got all this stolen property in there and he tries to give some of it to her, you know like, 'Here, open your car trunk and I'll give you these pants suits and stuff.' But she couldn't open the trunk 'cause she had a radio in there, the two-way system, so they split and evidently she turned that information over and they went back and busted him. I think she got a commendation for that and a hundred-and-fifty-buck bonus.

"But in Philly, I know her main assignment was check detail. I remember she used to complain about having to work for them all the time, tracking guys down who have so many different IDs and aliases, and then draft dodgers, stuff like that. Not really my bag.

"She was a real nut on shooting; she wanted to be an expert shot. She had a .38 that was kind of chrome-plated so that it wouldn't rust that much 'cause she used to wear it, you know, with dresses and stuff, she'd wear it inside and would get sweat and stuff on it. She'd been out here just before the crash, on her way back from Spokane or Seattle with the Saxbes, and I took her out, I thought Big Brother would take her out and help her . . . God. I went up there to the range and fortunately one of the guys that works for me is the best shot in the department. She got out there and started plugging away. I decided I wasn't even gonna shoot that day. My buddy shot and he had to do quite a bit to keep up with her. And boy, could she drink those Margueritas and go out and stomp it up. She had a cowboy hat and cowboy boots, it was her thing. She just dug it, wild time and wild music. She could dance up a storm.

"But," said Jim, "she had her theories on life—she was a real righteous person. There was good and bad and she was gonna be good and she didn't like bad guys. But she could play the role, she could play the sophisticated female, you know, the street hustler if she wanted. She could flex very easily, she could talk to anybody on their level."

Judy waves hello to a couple of rough-looking characters lounging around the bar. Eagles?

"Friends." A level gaze through the cigarette smoke. "Sheila and I might drive to a place in separate cars," Judy says, "but we always left together."

"I pictured Sheila," one of her Las Vegas friends said, "as the girl next door who grew up. She had gone to work very early in life and was probably striking-looking until she put on all that weight. Then she got mixed up with this married guy and that's where I think she got off the track. At times, she was very conservative, her Catholic upbringing I suppose, and then, occasionally, she'd toss it all to the wind. There was a weird dichotomy in her life, a struggle between what she thought she should and should not do. If she hadn't had this thing with that guy she probably never would have come to Las Vegas and might have just been the typical American girl."

Judy remembers the time that she flew east for a brief reunion with her friend. "One night we were going out, this was when she worked in Philadelphia and lived in New Jersey, and of course she wasn't going to take her gun with her, so she decided to hide it in the washing machine. The next morning, I turned it on and washed the gun. I heard it clunking and I knew what I'd done and Sheila knew it too. She took it out, I proceeded to do the washing, and she sat in the middle of the living-room floor with her gun-cleaning kit and she says, 'I want you to know that cleaning my gun was not on the schedule today.' She had to have the grips replaced. Then we went to this baseball game afterwards—the FBI clerks played this police department team—and Sheila told them what I'd done and they were all giving me orders: 'What do you charge for a machine gun? for a rifle?' I guess it was kind of a classic story after that."

After a year in Philadelphia, Sheila got orders for the New York office but was transferred to Washington within three months to join a guard detail surrounding then Attorney General William B. Saxbe and his wife, Dolly.

Peloquin: "I think the Saxbes' life style was such that she kind of enjoyed the assignment. They'd hop down to Acapulco or wherever it might be, Costa Rica or some bloody place . . . She seemed fond of Mrs. Saxbe herself, but I think it upset her that she . . . she didn't know whether this was helping her Bureau career or not. She'd like going shopping with Mrs. Saxbe or something and wasn't that concerned about somebody leaping out at her or throwing a bomb at her. It was just, it was kind of a bullshit assignment, but it was, at the same time, an honor for Sheila because the FBI was very careful whom they chose for it. When they chose her for that particular job, I got the impression that she was the outstanding woman they had—she wouldn't screw it up with the Attorney General. Just look at the male agents that they get, and I'm fairly familiar with them—they pick the outstanding ones because they don't want some boob sitting with the Attorney General. So, in a way, it was a great honor for her, but at the same time she was thinking, 'Jeez . . . I should be out getting my share of bank robbery interviews.'"

From March to December 1974, Sheila shadowed Dolly Saxbe across the continent, through stores, beauty parlors, official luncheons, restaurants, churches ("Sheila went to church with the Saxbes one Sunday," Judy says, "and Mrs. Saxbe almost died when she put ten dollars in the collection plate and took change"), and to the Attorney General's farm in Ohio, a lonely, quiet place where she took up painting and practiced her pool shots. Dolly Saxbe was to write:

I think of Sheila as a good friend and joyous companion. We were together constantly for nine delightful and interesting months. I never knew her to shrink from any situation that confronted us. Her enthusiasm for everything and anything was a pleasure to share. She approached life with energy, confidence and a sense of humor.

Her love for her country and her chosen service is seldom seen in the youth of today. She was competitive and she tackled each thing with one thought in mind, to be the best! I remember watching a TV program with her where a rookie cop turned in his badge because things got too tough. Sheila's comment was "They would have to kill me to get my badge."

I know that when she joined our squad the young men in our group were skeptical of a woman FBI agent. Without exception, they all came to admire and respect her. She never asked a favor because of sex. In fact, she could do as much or more than some because she wanted to excel. Many times she shouldered extra duties since they all had families and she was single. These things were always done cheerfully.

As a female, she had great qualities. She was a good cook and seamstress, had a wonderful way with and love for animals and plants. What a marvelous homemaker she would have been! I want to point this out, for one might think that with all these accomplishments as an athlete and a really tough agent she would have lacked femininity. Not so, she was a fantastic combination.

Sheila spent Thanksgiving in Mechanicsburg and on December 1, boarded TWA Flight 514 for the flight home and a few days off.

Jim Regan: "I may have worked a homicide or something the night before and I was sleeping late when this guy from the FBI came over to the house. I thought maybe he got a big new lead on some of the bank robberies we'd been working together. He said, 'I've got some bad news about Sheila for you. You'd better sit down.' And when you say that, you can't get much worse. Things were running through my head —did she get shot? Did something happen to the Saxbes? And then he laid it all out."

As a man who grew up on his father's stories of two wars and service in the Shanghai International Police, as a marine, a Vietnam veteran, as a cop, Detective Sergeant James Regan understands the premeditated violence of war and crime. "You work in homicide and you're pretty immune to death as far as emotional state or something like that." But as the only brother of a young woman killed in an airplane crash, "a useless, avoidable accident," he misses her more than he thought possible. "And I think since this happened— I don't know if I'm changing—but it seems like everything's, well . . . something's changing. I feel like I'd like to maybe leave Santa Barbara and leave law enforcement and just go do something else. But then again, maybe it's just a depression that I could get over. I feel different about my job and where I live [a twenty-minute ride from Flight Engineer Tom

Safranek's widow], and I feel like I just want to start all over somewhere else."

"I believe there's a purpose for everything and that somehow there must be a reason for this in God's great plan," says Judy Fensterman. "There's a reason for it—but I don't know . . . I don't know if you ever find out."

It is very late. A wind off the High Desert sweeps the neon avenues. The shows have broken and their stars sleep in pampered isolation. These are the hours of the night when high rollers cluster in special places to mutter incantations over green baize and rattling dice.

LAURA S. NOXON, 52, BODY BAG #89
ROBERT L. NOXON, 50, BODY BAG NONEXISTENT

Alice, the Noxon's daughter, a newly qualified practical nurse, lives in an unfinished "town-house" complex near Fort Benjamin Harrison in Indianapolis. She wears her short blonde hair in bangs and, in the popular press, might be described as "perky" and perhaps a little "chunky."

Alice speaks very rapidly in about three different octaves as she sits in her living room on a hot August morning happily remembering her father's little quirks and habits: "He was very methodical, very precise in everything he did. I guess it had to do with his engineering background. When he came for Thanksgiving, he had a new graph he'd worked on about life insurance, with his income starting from the very first job he had [at the Indianapolis Naval Avionics Facility where he eventually worked on the Polaris missile and other classified armaments], going on through his raises, and extended to what he thought he was going to be making until he retired [from government service and the Naval Air Systems Command in Crystal City, Virginia], and then the annuities he was going to get. He had it all mapped out . . . spooky. He just took out a new life insurance policy three days before he died. This is why he was mapping out his income, to figure out if he died how much life insurance he was going to need to provide for us in the way to which we had become accustomed. He took the policy out here in Indianapolis. There was a special agent he liked. It was a guy he knew and respected. Dad did that with a lot of things. Originally they were going to drive out, but Dad didn't feel like the hassle of a drive. He was tired. But when they were still talking about driving, Dad said he was going to bring back a color TV to the man he'd bought it from here, to get it fixed. And he always went to the same service station to get his car serviced."

Noxon, a balding man of fifty, with a strong nose and chin had also been working on a family tree. Alice remembers his telling her about it. "The Noxons came from England, before the Revolution, before any of that stuff. When the war broke out, they went to Canada—they were the first draft dodgers—and after that moved to New York, upstate, and founded the Noxon broom company. Then they spread over the country and eventually most of them wound up in the Midwest. There's a place in Delaware called Noxontown with the Noxontown pond. A relative of ours built a grist mill in the middle of the pond. It was written up in the *National Geographic*. Oh, that's what Dad wanted to do when he retired, take pictures for the *National Geographic*. But Dad hadn't gotten to that point yet; when he died he was still working on the New York part. The Noxons had been married for nearly twenty-three years when they died, leaving three children: Alice, her brother Jim, and Sue, their stepsister from Mrs. Noxon's previous marriage.

"When Dad got as high up as he could working here in Indianapolis, but wasn't near retirement, he had a chance to go to Crystal City. He went in June of 1970, and we followed in December. Or rather, I was a senior in high school so I stayed here for three months to finish—things came up and I didn't finish school; being alone and seventeen, it didn't work out like I thought, so I went out there and finished.

"I lived in Virginia for two years, but decided I didn't really like it and that I was a midwesterner, so I came back and enrolled in an LPN [Licensed Practical Nurse] school. My Mom was a nurse, an RN [Registered Nurse]. While we were growing up, she didn't work, but in about 'sixty-five she went back... It was starting to be college time for the three children and money was needed."

The first trip Alice remembers her father taking abroad was in 1968 when he went to "the Philippines, Japan, and Hong Kong on an aircraft carrier testing out some radar device. He took jillions of slides and had a travelogue all worked up on that trip which he showed *everybody*. He had suits made for him in Hong Kong, had one made for Mom and brought me back a fall and some stuff which didn't fit. He'd used a color picture to choose the fall and it came out too light." She laughs now at what must have been a disappointment, presents from far away that didn't work out. Presents from Dad, who had come out to Indianapolis for Thanksgiving, years later, "to meet the guy I was supposed to get married to.

"They'd come out to meet him and about three days before, I called him and asked if he was going to be here. He [pulled the] I-think-I'll-go-back-to-the-ex-wife trick. I thought of calling Mom and Dad and saying, 'If you don't want to come out, you don't have to.' But until the last minute I was still hoping he'd change his mind, you know, look what you're missing, ha!" A small, lady-like snort. "But he didn't, so... boom. Two sock-it-to-you's right at the same time.

"My parents and then him... We hadn't decided on a definite date. I guess I wasn't really engaged, although I thought I was." A hollow laugh. "Funny the way those things turn out. I saw him twice in the past month, but after the crash, nothing—not a phone call; not even an American greeting card, you know, sayings for every occasion?"

But Thanksgiving weekend was "just kind of a nice, re-laxed, happy weekend, nothing heavy; we just kind of got back to each other." Mr. Noxon had Alice's roommate go out and buy a toilet bowl cover, "some squirt for windshield wipers and he bought me snow tires. He went out and did it, wanted me to have the best."

And on Sunday, after driving her parents to the airport and rushing home to get back into a warm bed, the phone ringing and her roommate's voice on the other end telling her the plane had crashed and not to panic.

"My first immediate reaction was *no*. NO, absolutely not. Okay, the plane may have crashed, but they're all right, nothing happened to them. So I called my brother Jim— he was then in Virginia [unable to come for Thanksgiving because he'd gotten a free-lance assignment to take wedding pictures]—and said, 'Have you heard anything about this?' He said, 'No, I've been listening to records all day.' I said, 'See what you can find out; you're closer to it than I am.' So he called around and tried to get to the crash site, but they wouldn't let him. Then he called back and said there were different reports. One radio station said there were no sur-vivors, another said sixty-six people were killed and yes, that there were survivors. I of course latched onto that.

"TWA kept transferring me back and forth on the phone so that with everybody that I talked to I had to go through the whole story again, and nobody could tell me anything. They said, 'We've never had this happen before.' I said, 'Fine, I haven't either.'

"About ten that night, just before it came on television, TWA finally called back for identification purposes. I did great on Mom; Dad I wasn't too sure because he went for grays and browns.

"The second call I got that day from someone who wasn't a relative came from the newspaper, and the guy said, 'Is this A. Noxon?' I said, 'Yeah.' He asked, 'Do you know an R. Noxon and an L. Noxon?' I said, 'Who?' and just played it like I didn't know what the hell he was talking about and he said, 'Well, they were on the plane that crashed.' I hung up and immediately called the paper back and gave them hell. What right have they to barge in on me like that? It happened to my brother worse. He was there alone [in the Noxon's house in Crystal City] with a buddy, and a TV crew came to the house, came into the living room, set up cameras, lights, everything. And Jim doesn't know what's going on, and they're shoving a mike at him . . . It's just unbelievable.

"The second thought I had was to get to Virginia, to get to my brother, who was then nineteen. I wanted to be with him. I finally talked to someone at TWA. The mistake I made was not calling Reservations. I talked to this poor guy who was handling this, and asked if he could get me on a plane. All I kept getting from him was, 'You can't go to the crash site.' I kept screaming at him that I didn't want to go to the crash site, that I didn't want to see it, that I had a brother out there who was alone. They said, 'Well, we'll see what we can do to get you on the next plane, and you won't have to pay.' I said, 'I have an aunt and uncle who are going to go with me because I can't face this alone.' They said, 'Fine, we'll fly them free, too.'

"Well, I never heard anything for the rest of the night. In the morning I called back, got a different guy, and asked what had been done about my reservations. He said he didn't know what I was talking about, and again the 'You can't go to the crash site' story. I mentioned my aunt and uncle, and he said, 'For free? Huh, I don't know who told you that, but sorry, no way.' "

Alice finally flew to Washington on a round-trip ticket her father had bought her so the whole family could be together for Christmas. After the memorial services there (there could be no funeral yet since neither of her parents had yet been identified), Alice and her second cousin who had taken over the duty of administering the estate went "through the house stem to stern looking for a will." They never found one. "It still bugs me because Dad was so methodical. We went through check stubs trying to find a lawyer's name, we went to the only lawyer I know they ever went to—he'd handled the mortgage. Nothing." Then Alice came back to Indianapolis for memorial services here. Again she paid her own way. "I was invited to a friend's house for Christmas Eve, but couldn't go. I spent Christmas here—sat around and cried."

The day after Christmas, she and her roommate drove to Virginia to collect all the plants in the Noxon house. She points to the philodendron filling a corner in her small dining room. "See that? It's twenty-two years old—today. It's my birthday, yeah. When Mom was carrying me, she took a slip from the doctor's office, and that's it. Everytime that thing looks peaked, I take my pulse. I figure as long as I can keep it all right, I'm doing fine."

Clearing out and readying her parents' house for sale almost undid her, she says. "It took three weeks; [I] had to go through the clothes and the drawers. It was just unbelievable. The first three days I went out there by myself and I'd just walk in and . . . dissolve into tears. I just kind of walked around, put things back in cabinets. I'd go into one room and do one or two things and then go into the next one . . . I was totally disoriented. I didn't know what I could possibly pack or accomplish.

"After the third day, Suzie [the wife of a close friend and ex-neighbor of the Noxons] saw what was going on and she came out with me and several of the neighbors and got boxes and said, 'This has to be done; let's start.' I had someone else go through the clothes; I couldn't do that. I sent them all to Goodwill. Some that were especially good I sent to a thrift shop."

Alice guesses that her brother and stepsister haven't quite "turned out the way Mom and Dad wanted them to," and that she now finds herself "responding to things even more the way [our parents] would have. I'm in the process of finding myself again. I get to the point, especially now that the crisis is over, things have taken their course, everybody is doing their own little thing, and I look around and say, 'When is it my turn to be weak. I need somebody whose shoulder I can cry on.'. . . When I heard about the crash, I immediately wanted to smother everybody involved so it wouldn't hurt them, and I didn't want them to worry about me. So I immediately took on my little Rock-of-Gibraltar stand. It hurts, it hurts bad, but I'm making it fine, don't worry about me." She smiles and pours more coffee, and adds, "I can't count the times I've sat down and thought, 'Why didn't I tell them I loved them just before they left?' "—as they walked out the door, her father in his grays and browns, her mother carrying a copy of *Airport*.

Her voice changes, takes on a brittle, slightly defiant timbre as she talks about the logistics of her parents death and of the certification of that death. Though Alice's mother had been identified the week before Christmas, there still wasn't any word on her father when the NTSB hearings began two months later. "In the meantime," she says, "everything is tied up. You can't settle an estate if you don't have a death certificate. That's the first thing everybody asks for, logically enough. The insurance companies wouldn't accept a clipping out of the newspaper. In the meantime, I haven't worked, nor has my brother." Money had to be borrowed from relatives, and the heartless idiocy of the whole situation was further brought into focus by the question of compensation. "I was approached by an attorney from TWA, or somebody who worked for a firm that represented TWA here in Indianapolis, and he was going around to everybody here telling them they didn't need to hire a lawyer or to file a suit, that everybody would be settled with equally and that we didn't need to go to all that added expense. That an attorney was just going to take out blah blah blah per cent. The man was very nice and very convincing and if Les [her second cousin and administrator of the estate] hadn't been around, I probably would not have hired a lawyer. I believed the man. Why would he lie, you know? Dumb question. Why would he lie . . . He was just trying to do his job for TWA, trying to keep all the hassles off." Newfound cynicism.

"Les was approached by F. Lee Bailey—bleep." Alice doesn't actually say words like that. "Took Les out to lunch and the whole schmear, had the contract all ready and everything. I thought, oh well, F. Lee Bailey . . . I'd heard the name on the news, a very famous lawyer, been on the 'Tonight Show.'" Apparently Bailey was no more convincing to the Noxon clan than he was a few months later when he announced, in a televised Indianapolis news conference, that he was filing a class action suit on behalf of all the victims' relatives. A motion was filed by a firm of New York aviation attorneys, in U.S. District Court in Alexandria to invalidate the class action, and Bailey's interests in the case were somewhat diminished when Judge Albert Bryan dismissed his class action suit as inappropriate.

But Robert Noxon had still not been identified, and the FBI asked Alice's brother for some toenail clippings or hair samples of their father's. Anger mixed with incredulity as Alice recounts hearing of the request. "Dad had his razor on the plane. What is all this crap!" No "bleeps" this time.

"Finally we found out [his blood type] was O-Negative. We got word to the coroner. By now the body was so decayed they couldn't measure it." But her cousin Les had apparently been told that ninety-one of the ninety-two people on the flight had been identified, and Alice remembers saying to herself, "Okay, if you have ninety-one and there were ninety-two on the plane, that's Dad. He's not around."

What Alice didn't know—because her roommate had tactfully hidden any pictures of the crash site published in the newspapers—was that the impact of the crash and the subsequent fires had shredded and burned many of the victims beyond any semblance of recognition and that, although the FBI had many of the passengers' fingerprints on file and had asked for and gotten most of the others'—including Robert Noxon's—there was often nothing left to be printed.

In despair, Alice's cousin leaked the story of the missing body to some of the reporters covering the NTSB hearing. A television station picked up on it, filmed a story, but shortly called Les to say they couldn't air the piece because TWA had told them all the bodies were accounted for. Somehow, perhaps by extraordinary coincidence, Noxon's death certificate was signed the next day. "So," Alice says shaking her head, "still in the back of my mind, there's the thought, Okay, they did that under stress. Are they really sure that was my father's body?"

The answer is no. All that the pathologists working up at the Bluemont School and then down in their Richmond Labs could determine was that "body part #7 consisting of a segment of male vertebral column of blood type "O" probably represents Robert Lowell Noxon."

RODNEY E. STEELE, 37, BODY BAG #X-1

"He ran much faster than he had expected—laughing and joking as he passed people he knew along the way. He was having a good time. 'I may be running too fast,' he said when he passed me, 'and I may see you later, but I hope not.'"

—GRAHAM HOUSTON

Wellesley Square... Center Street, Newton... Heartbreak Hill... his pain threshold long since reached and ignored, Rod Steele never slowed and Graham Houston, his friend, never caught up. Stride by burning stride, running with that peculiar, terribly erect style of his, Steele pounded past the pack strung along the 26.2 miles of New England roads and city streets known as the Boston Marathon. Neil Cusack of Limerick won that year in 2:13:39, and did so with ease. But when Rodney E. Steele crossed the finish line on April 15, 1974—though he was 33 minutes off the Irishman's pace and though his legs ached mightily—he had just run the best race of his life, placing 263 in a field of nearly 2000.

No full-time athlete, Steele was a bureaucrat among the multitude of Washington bureaucrats—an economist for the CIA and then for the Department of Agriculture, drawing his pay to help support a wife and a child who, at the time of his death, no longer lived with him.

Houston has a pretty fair idea of why his friend ran; as the attorney for Steele's estate he had had to open the files, read the letters. Tactfully, he says only that "We all run for different reasons. It's a release, it's an escape . . . a mind-cleansing experience."

Roughly a fifth of its way across America, Interstate 70 neatly cuts Arlo Steele's land in half. Gazing south toward the rumble of traffic from the knoll on which his farmhouse sits, Arlo says, "The highway comin' through took about seven or eight acres. We have about thirty-three left, twenty under cultivation." The young corn lies low and green and the bottom land, loamy and giving underfoot, smells of promises. It is early June and it has been raining in Ohio. A westerly breeze carries hot asphalt fumes from the plant down at the end of Lower Valley Pike. "Not much you can do about that." Arlo stomps mud from his work shoes and goes into the house. Sunday dinner of pot roast simmers on the stove; it is the Steeles' forty-fifth wedding anniversary.

"We're both sixty-six now, I'm going to be sixty-seven this fall and we're wondering how much longer we'll be able to keep things going," Esther Steele says. "We haven't any plans."

"We farmed this farm since when the boys were little," Arlo adds.

"Now," says his wife, "we wonder how we ever did it, you know, working full time and then coming home . . ."

Discharged from the air corps at the end of the war, Arlo bought his forty acres in 1946. ("Raising boys, there's no place like a farm to raise them.") For the next nineteen years, he worked as a mechanical engineer at plants in neighboring Springfield, Dayton, and Urbana. Retired now, he and Esther till their twenty acres of corn alone. "They keep me in physical condition," he says.

"Well," Esther allows, "we have more modern equipment than we did when the boys were home."

A farmer's dream brood, their four sons followed each other through Tecumseh High School in a twelve-year span. Esther relates that along with their farm chores—the Steeles then kept a few head of cattle and a few dozen hogs—the boys all held summer jobs "when there were some available. Both Rod and Larry worked for the cement plant and they said that's what helped them decide they wanted to go to college—after working in a cement plant." She smiles, herself no stranger to manual labor.

The Steele sons succeeded each other at Ohio State University, and Rod, third in line, graduated in 1961 with a degree in economics. Recruited as an Intelligence Officer Industrial Economist by the CIA, he moved to Washington, D.C., only returned to Columbus a year later to marry his college belle, Bonnie, and then headed back East with her.

"I know somebody else in this neighborhood who is inside the CIA. *His* mother would always talk about the things he told, but Rod would never mention a thing."

The younger Steeles moved into the ultimate suburban community of Reston near Dulles Airport; Rod joined the D.C. Roadrunners and Bonnie sporadically attended graduate school. They conceived a child and on March 17, 1971, rejoiced at Cathy's birth. Eight months later Rod drove home from CIA headquarters in McLean on a Friday night to find his wife packing.

"He thought they were getting along so much better. He was real surprised when she said she was going to leave the next day," taking the baby with her. "He hated to break the news—it was right around Thanksgiving, a difficult time to break news like that to us. And they were pretty bitter toward each other for a while, but I think that was the finest thing that he could do, that he did everything he could to make life easier for his daughter. That was one thing he said, he was always hoping to be around to..." Mrs. Steele stops, looks at her husband. Slumped in a rocking chair, he stares at his hands. Apparently, Bonnie did not always think of Rod the way his parents did.

"... he wanted to live," Mrs. Steele whispers, "to show Cathy his daughter, that he wasn't a heel."

Behind his glasses, Arlo Steele cries silently.

With his wife and child in Ohio and his job at the CIA about to end, Rod had moved into Washington, secured a position at the Department of Agriculture, still as an economist ("He was so open about his work with *them*"), and trained for the Boston Marathon. His *joie de vivre* grew apace with his stamina. He met a girl, experimented with gourmet cooking—"the runner's reward," one of his marathoner friends says—and served as president of the D.C. Roadrunners. Then, on a sunny April afternoon in Boston it all clicked.

When he came home for Thanksgiving in 1974 he was happy, says Esther. "I think it showed up in his running and the marathon was a big thing in his life. He was so proud of his time, two hours and forty-five minutes and some seconds I believe it was ... Of course we got the trophies."

But he didn't run in those four days back at the farm on Lower Valley Pike. "He would think about it," his mother says, "but his little girl was here with him and it seemed he wanted to spend every minute with her. She was four in March. What I keep thinking about was he just couldn't stand to be away from her, and he made a tour of the farm, he and his brother and children, just walked around, talked about things they had done as kids.

"He went to the Springfield Mall to buy Christmas presents for Cathy. He wrapped the gifts and left them here, planning to come back at Christmas time and give them to her ..." A pair of F-104 fighters on approach to Wright Patterson Air Force Base shrieks over the house, rattling windows. The two old people wince.

Mrs. Steele says she has had to "reconcile" herself to the crash and that she thinks "of all the parents who have lost sons in war, and that makes it a little bit easier for me, thinking at least it didn't happen that way."

There is a long, long pause.

"It sure does seem kind of useless," says her husband, looking up from his hands.

"Yes," Esther says, "when you think about it, it could have been prevented, you know."

They were never called by TWA, they say. The way they got confirmation of their son's death was on the Sunday night eleven o'clock news.

Mrs. Steele says, "I shed more tears now than I did at the time. I think, you know, it just didn't seem real to me, and all through the service you know, since the casket was closed, I don't think I cried at all, but now, just any little thing . . . I think Memorial Day weekend was the hardest on us because his birthday was on the twenty-seventh, so between Memorial Day and his birthday why, it was an awful time . . . That's why we miss him so much now, because of the holiday seasons, you know, he'd always been with us. Thanksgiving, Christmas . . ."

Cathy spent the holidays after Rod's death with her grandparents. "For a while we just couldn't stand to see her open those things," Esther says. "But 'cause she was so sweet about it, it wasn't nearly as difficult as we had thought it would be. We do wonder how much she will remember of him. I think that sort of worries us, 'cause now the memory is fresh, but how long will that be with her? So, we have made up some albums of Rod from his youngest years on up. When she gets older, maybe eighteen or twenty, she might really appreciate having them."

The crash?

"Well, I don't—we didn't know if she had been told or not. At Christmas time she said, 'Daddy can't be here,' and that's all she said. And then, when she came at Easter, why, she said, 'Daddy died.' Now whether she knows what that means or not, I don't know."

Mrs. Steele goes to look after her roast. Her husband and I walk outside; he points to his three apple trees—one golden delicious, one crab, and one "just good eatin' apples."

"Planted the trees," he says, "so they'd have some fruit when the boys grew up. Took them fifteen years though . . . By then they'd all more or less left."

GERALD S. ("BUCK") SEYMOUR, 19, BODY BAG #88

Gerald was not a marine, but a simple private in the U.S. Army earning $383 a month who, when he died, was headed for the old red bricks of Fort Belvoir, Virginia, down by Mount Vernon. He had been inducted into the Army on July 24, 1974, the day after his nineteenth birthday. His honorable discharge, posthumously awarded, arrived in the mail five months later addressed to James and Viola Seymour of Terre Haute, Indiana. The Army had taught Gerald how to be a typesetter and had planned to ship him to West Germany after Christmas. He would have had to leave his ten-speed bicycle behind.

The potholes on Thompson Street where the Seymours live at Number 2606 are the kind which are seldom filled except in summer, and then only by the dust which hangs thick and gray in the wake of exhausted Chevies and Fords. But it is September now and raining. A Pepsi-Cola can and a blue-and-white-striped engineer's cap lie squashed into the sodden grass by the porch.

James Seymour opens the door before the echo of the second knock has died. His wife, barefoot and bare-armed in a loose, thin, red and yellow dress, sits on the couch facing the innards of the house. It is dark because of the rain but no lights are turned on. Seymour has gaps between his teeth and the words come out a little slurred and in a gentle rush as though everything has to be said in the same breath. Conception, birth, life, and death in one lungful. Reluctantly, he pauses for oxygen.

". . . well anyway, we took him up there to the bus station on Sunday morning. He took the bus to Indianapolis to catch that plane out. See, them big planes they don't land out here at home airport. See, we got an airport here in Terre Haute but it's just for smaller planes. These big jobs, Eastern Airlines and TWA and all, they go in and out of Indianapolis." *In-ya-napolis.*

"The bus was late. I think he had to check in for the bus at five, but the bus hadn't come and we had to leave 'cause I had to get back, I had to go to work." (In the boiler room of the Western Paper Company—"See, they got to have steam to dry the paper, see they make this paper out of wood and of course then they gotta have steam to cook the wood. They call it the fire department. Been there twenty-three years.")

Viola: "He said he'd call me. Usually he calls me from Indianapolis. But he said, 'I won't call you at Indianapolis. I'll wait till I get to Virginia and then I'll call you.' But he never did call. And then we were sittin' watchin' TV and we heard it."

James: "Yup . . . well, I was at work and of course they called me over there, at work, and told me about it, you know . . . Well anyway (deep breath) I had just about an hour to go, so I didn't just take right off and come home, you know, 'cause I figured well, there wasn't nothin' I could do."

Viola nods. They look at each other.

James: "And so anyway, when I come home we tried to get hold of TWA over in Indianapolis, you know, was having a job of getting through and so . . . I even . . . on the news . . .

we watched . . . they showed the picture on the channel, over there in Indianapolis, you know, of the plane and everything So anyway, they said there were no survivors in it."

Viola: "Then the army, they came. They was supposed to be the ones to come and tell us. Of course, we already knew. They wanted to know how come, how come we found it out, and I said, 'Well they give the names right there on the TV. In great big letters.' "

But the Seymours like the army and what it did for Gerald, which they pronounce with a hard "G." It was the army which gave him his chance to break out of Terre Haute, learn a trade, and hold a steady job—things which, they quickly point out, none of six brothers and sisters (three from Viola's former marriage) ever managed to do. It even gave him a pair of eyeglasses which he hadn't known he needed.

James: "He had several things he tried to take up. First, he wanted to be an artist, drawings, you know and then . . ."

". . . then, when he was young, he wanted to be a preacher. But he got over it." Viola laughs.

James: "The way things was and jobs hard to get around here 'cause they laying back—we'd been on strike for a month over there [at the paper plant] and just got back to working when this happened—I told him that, Thanksgiving morning when we went over there [to the bus depot] to pick him up and bring him home for Thanksgiving dinner, I told him everyplace was laying off, just cutting down to skin and bones, I told him, I says, 'Boy, you sure lucky gettin' into what you got.' I said, 'The way things is now, you got into something that's gonna help build your life up.' "

Viola: "He said, 'I'm going to get to the top. I'm going to show you and Dad that I can make it in this world and I don't have to have your help. I'm going to show them boys up.' " She wrinkles her toes on the carpet—"He was the best of 'em"—and illustrates by sketching the employment record of her other three sons: Harold (Gerald's twin), nineteen, grass cutter and part-time janitor; James, Jr., twenty-one, grass cutter, high-school dropout; Clarence, twenty-eight, grass cutter. Though obviously bemused by the other boys, disappointed even, the Seymours are tolerant of their fruitless casting about in search of work, of themselves. Tolerant despite the tight food budget, the medical bills, despite a farming background where everybody pulled their weight from dawn to dusk.

James, Jr., ambles in from the kitchen, a big soft fellow with uncut curly hair. He plunks himself down in a chair.

Asked what his plans are he shrugs and says, "I don't know."

"He's gonna sit there until that chair goes to the basement," says Viola.

"Really," her son says reprovingly. "I just haven't figured it out yet."

"It's about time. You're twenty-one years old. You ought to know . . . He has a little paper route," she says turning from him. "But that only takes about an hour. Then he sets and watches TV. He knows everything that's on TV. He don't need a *program*, 'cause he knows it all."

She pauses. Her son grins at her, proud of his mastery of daytime television programming the way some boys show off their knowledge of railroad schedules. She starts in again, almost as if playing a too-familiar game. "Yeah, but you gotta be thinking about what you're gonna do with your life. You know what Buck told ya, that he's gonna kick your hide in unless you do something . . ."

Silence.

James: "Buck was a boy never was . . . If he wasn't doing something, he was on the go. He couldn't sit in that chair fifteen minutes right there and watch that TV."

Viola: "He'd take bicycles out. He had a speedometer on his bike and boy he'd just take it everywhere. Yeah, he had a brand new ten-speed, then he had this red bike, he kept it all shined up, there wasn't a scratch on it. Well, after he passed away, I give it to one of the other boys. . . . They don't look like the same bicycles. No, they don't."

James: "He's the type of boy . . . all these boys nowadays you tell 'em to do something, you gotta holler a hundred and one times and they won't get up and do it. He's the boy you tell him once what to do and he'd do it. Like he had two jobs here, we got two dogs. He took care of the dogs. And we got a fish bowl settin' right over there. And he'd always kept it clean. We did have a lot of fish in there but actually, since he passed away, the fish started dying."

Viola continues: "Well, we got a big algae-eater. Oh God, I'll tell you something, he should have been dead a long time ago, that algae-eater. But he's still alive, and Buck says, 'Now if that algae-eater dies while I'm in the service, I want you to keep it till I come home, because I want to experiment with it.'"

Unconsciously, the Seymours slip in and out of the present tense when talking about the dead boy.

They are considering TWA's settlement offer of $50,000, up from the initial $32,000, but haven't yet decided whether or not to accept it. Not that they seriously plan on suing for more in court, just that in accepting the money they also have to accept the fact that Buck is really gone.

In Viola's mind there is still hope. "We was getting some phony calls there after this happened," she says. "I got a phone call from someone, a young guy, that said, 'Is this Viola?' And I said yes. And he said, 'This is your son Gerald.' And I knew it had to be phony because he wouldn't say 'Is this Viola?' He'd say, 'Mom, Mom, this is Gerald,' or Buck or something like that. He wouldn't say that . . . I just couldn't think who'd do something like that. And I knew, see, I knew it wasn't him 'cause I knew it wasn't his voice. And then I had this in my head all the time that he wasn't dead in the first place, see. I thought, 'Well, you know, I just had to go by their word, you know.' And I never did believe he was dead and then that phone call come . . ."

Mr. Seymour looks at her, says nothing.

"I didn't believe it. And that's just the way I felt. I did

not believe it and I had a hard time because they wouldn't let me see in the casket when they brought it back. I wanted to see that it was really him in there. To me it seems like he's still in the service. They didn't open the casket and I could not believe he was in there. And the army man said, now the government wouldn't send an empty box back here 'cause they have paid for it . . . They wouldn't send that back empty without somebody in it. So, I still can't understand why they didn't let me see.

"He's not dead, because I didn't see him . . ."

TIM HARLAN, 29, BODY BAG #14

As he approached Cincinnati around noon, driving south on I-74, Lee Giles, news director for WISH-TV in Indianapolis, idly listened to the chatter coming in over his car radio. It was the Sunday after Thanksgiving and he was en route to collect his wife and children in Kentucky. An announcer broke in with a bulletin. A TWA plane had crashed. He turned up the volume to catch the details. They came in piecemeal— . . . hit a mountain in bad weather . . . ninety-two aboard . . . stopover in Columbus, left from Indianapolis.

"I thought, Oh my gosh, there were lots of congressmen that flew on that flight. I got to my mother-in-law's house and called the station to see what was going on, to tell them where I was and see if things were under control. They were on the story and had a crew out at the airport trying to get some local stuff. Bruce Childs was the guy I was talking to—he's co-anchor on the Sunday news—and he was working on a story and they were snowed under. I asked him whether there were any prominent people that he knew of on the plane. He said no, but said, 'Hey, that guy that was in here yesterday was on it.' "

"I thought, What guy? And he said, 'Ah, you know, that friend of yours.' And I said, 'Not Tim—' and he said, 'Yeah.' "

WISH-TV had a short special on the six o'clock evening news the next day. Anchorman Mike Ahern faced the studio cameras, straightened his tie, nodded, paused, and said, "The passenger list of that ill-fated TWA jetliner yesterday carried the name of Tim Harlan. That name won't mean much to most people, but to his family, his friends, and those of us who worked with Tim here at Channel 8, it's a loss deeply felt. Tim was a news photographer here and a good one. Tim was a photo-journalist in the best tradition of the word. A photographer with a keen sense of what is news and how to capture the essence of what the news is with his camera. But more than that, his was a restless talent. Tim was always looking for new ways to work with film. Never satisfied with what others had done, he constantly went exploring. At times his imagination, his creative energy had a life of its own. Tim simply followed it wherever it took him. At last it took him to Washington and several successful years as an independent film producer. The irony of it is Tim told me just a few hours before his plane took off that he was hoping to return to Indianapolis soon. This was his home and he missed it. With Tim Harlan's death yesterday, the film industry, the whole business of communications, lost a rare creative talent and we, here at Channel 8, lost a good friend . . ." FADEOUT.

Up on Weather Mountain the pathologists and FBI fingerprint specialists matched names with mangled flesh. In Indianapolis, Alice Lindenberg, Tim's mother, tried to stop crying. A few miles away his wife, Ellen, seven and a half months pregnant with their second child, resumed her switching of channels in an unflinching quest for information. Across town in the Ponderosa mobile home park Tim's father, Ken Harlan, had missed Ahern's electronic eulogy as he too dialed his television set from channel to channel trying to understand, wanting to be told, to be given one good reason why ninety-two people had had to die at 11:09 the previous morning. There were no explanations—just a long list of names, the roll call of the dead: Goldblatt . . . Griffin . . . Harlan . . . Hasenkampf . . .

The long, black train winds through green Appalachian hills. You see it from above, from the side, coming at you. It's just a coal train but from the angles and the impressionistic fuzziness of the color you expect to hear bluegrass chords and have the train lead the camera to a boy and a girl holding hands on a bridge. Cut to a switching yard . . . the pace quickens . . . closeup of two cast-iron couplings clashing, recoiling, holding firm. Water splashes off the colliding metal. END OF REEL.

The last bit of celluloid flaps through the projector. Steve Eschelman turns up the lights and rewinds. "Tim loved shooting this thing for the West Virginia coal people," he says. "He loved trains and got to crawl all over them. We shot the couplings coming together in slow motion to catch the water, which was his idea—he squirted it on—to add vertical motion to the shot."

It's the end of February in Washington, three months after the crash and exactly a week after the end of the National Transportation Safety Board's hearings. Eschelman, Tim Harlan's friend and colleague at Creative Arts Studios gladly runs Tim's old footage. A private little tribute in a cluttered cutting room.

"Tim had a rare identity level with that one," Eschelman says, putting the film away. "You know—because we make industrial films here, the satisfaction we get, the satisfaction Tim got, is personal really, not public."

Walter King is president of Creative Arts. Curly black hair, gray flannel suit, pink button-down shirt, almost but not quite Ivy League. A friend and associate of both Tim and his older brother David. "We make educational, safety, and training films," he says, "mostly for private distribution. Tim produced and directed, Dave was the more creative side of the team, the writer . . ."

Dave Harlan comes into work. Between phone calls, he describes the partnership they had. "Tim felt he'd gone as far as he could in the news business. His choices were, one, to go up the TV station chain of command, out in Denver, or two, to get into the creative side of film. He approached me during Christmas in 1972 in Indianapolis and asked about joining me [Dave ran his own production company, called Palladium]. He came to D.C. around New Year's in 1973 and we talked, and agreed. He sold his house in Denver and came over. We were all doing well, then, suddenly, the business copped out. Walter King, who used to be my attorney, had bought Creative Arts and saw the potential to stabilize the situation by putting the two groups together. Tim first came over to Creative [he was named vice-president in charge of motion pictures in January 1974] and then Palladium and Creative Arts pooled resources. Things started picking up, you know, one overhead, one set of buildings, one staff . . .

"Tim was a fantastic administrator, a stickler [for] detail. If there was a dollar to be saved, he'd do it. We had an intuitive understanding about work though. It saved a lot of hassles."

The two brothers also had a pretty fair understanding about how to cope with holidays. "Tim said he'd cover Indianapolis for us Thanksgiving so we could go to see [my wife's] family in Texas and then go on to see some friends in Juárez. Before the accident—which we heard about on Mexican television—we'd all decided not to go home for the holidays anymore, to get the parents to come to Washington. But our stepfather has a big Swedish thing about holidays. So Tim and Ellen and Nicole went . . ."

Indianapolis, September 18, 1975. The National League football players settle their strike, the FBI arrests Patty Hearst, and Ellen Harlan says sure, she'd love to talk about Tim. How about lunch tomorrow?

Lunch. It's Ladies' Day at the shopping mall and there's a twittering line for tables at the Magic Pan. The mall itself, in the northeast corner of the Indianapolis Beltway, is like shopping centers everywhere: plastic, faceless, a cheap simulation of opulence designed as a backdrop for the clutter of consumer goods.

We recognize each other almost immediately. Ellen is thin, with slightly disarrayed brown hair and light eyes. She speaks in a high, clear voice and laughs easily. We drink bloody marys and eat thick, lukewarm crêpes. She's brave. There are questions I want to ask but don't, yet.

She and Tim met in high school. She was dating someone else. He and his gang were "wild." She never thought . . . Well, they married. He worked as a cameraman for a local Indianapolis television station, covered the usual array of small-time politicians, four-alarm fires, head-on car collisions. Offered a job in Denver, they moved. Named head cameraman, Tim built a house. In Indianapolis for Christmas 1971, the two brothers discussed their old dream of making pictures together. "They'd always talked about having a business of their own. We wouldn't have left [Denver] at that time if Tim hadn't gotten this one new boss. News directors were always coming and going and this guy was a bastard. And . . . well, being as young as we were we probably should have stayed and hacked it out a bit more, but we didn't. Oh, we loved Denver, loved our new house; we climbed mountains, hiked really, had picnics by streams, had good friends there . . . But I think basically this new man was kind of a klutz and if it hadn't been for him, we wouldn't have left—then."

Tim was not unhappy about getting out of the news business; he felt mostly relieved not to have to see "all those ugly things" any more. "You can either get callous and shut your eyes and continue, or you can say, 'I'm fed up with looking at all of this.'" Tim decided to move back East and join his older brother's struggle to make a go of Palladium Independent Productions. And Ellen became pregnant "the minute we walked in the door to Washington," and gave birth to their first daughter, Nicole, in January 1973. Life was "financially shaky" during those first years in the capital. "We couldn't really just go bananas doing things." It isn't a complaint, just a statement of fact. But when, in early 1974, Walter King offered Tim a job at Creative Arts, he gratefully accepted it.

Ellen is not sure how Tim and his brother felt about Tim's leaving. "I mean he and David loved each other, but the business at that time couldn't support two families," Ellen says. "And it was David's from the beginning. We felt, well, you both can't starve."

Though the new job at Creative Arts provided Tim with a steady income, it also meant his working long and irregular hours, just the conditions he had hoped to escape when he left television. "This bothered us," Ellen says, just as it had frustrated and separated them while they were still attending Butler College in Indianapolis. She says Tim would come home from weeks out on location mentally drained, and that in the beginning, she'd be so depressed by his absences that she "cried the whole time." Finally they both realized that they'd both "had it with working twenty-four hours a day, and not having a lot of time together. We were very fortunate that we realized it. One year we made a hell of a lot of money, and the next . . . nothing. We got to realize that money is not everything. It was a fantastic thing to realize. Why should you beat your head against the wall? You should do what you want to do and be what you are. I'm a person very oriented toward [being] happy. Everybody should be happy. Everybody should do what they want. I told him, 'If you don't like it, quit. Let's do something that makes us happy.' I wanted him to go [back] to Denver and sit up in Evergreen, this little town [on the edge of the Arapaho National Forest]—we have this favorite bar up there—and live in the mountains, and if he got a job and worked a few days, fine."

But Tim was not that kind of man. "See, that's why we

didn't move back to Denver—he didn't want his old job back. That would be an ego come-down and he was . . . better than that. He didn't have to do that."

They compromised. "We talked about a lot of ideas, we talked about a lot of crazy ideas . . . but, we talked together and then, finally, he always made the right decision." Their decision in this case was to return to Indianapolis. "One of the several reasons is we're very close as a family and felt, 'We're still young but our families aren't going to live forever.' We wanted to be with them and we thought grandparents were very important. Plus we wanted to do what we wanted to do instead of having to be, you know, at work and at the grindstone."

Around noon on the day before the crash, Tim had called Channel 8 and spoken to Lee Giles who remembers, "He wanted to have lunch with Mike [Ahern], but he wasn't in so Tim asked me to go to lunch with him. I said I couldn't leave the office but suggested that he come in." Giles says he's sorry that "none of us could give him a whole lot of time. He obviously wanted very badly to go to lunch with someone, sort of 'Gee, I'm back in town and no one wants to go out with me' sort of thing"—which wasn't the case. Tim wasn't working; they were.

But he drove down to WISH-TV's sleek new building on Meridian Street, settling for sandwiches at Giles's desk and a chat about their respective children. When Mike Ahern finally came, Tim ran some of the pictures he'd made and lugged, along with the baby food, all the way from Washington. "He was proud of them, obviously," Ahern says, "and wanted us to see what he'd been up to." Ahern had worked with Tim on one of Tim's last assignments for Channel 8, a special on Neil Armstrong just before the first moon shot. He promised Tim he'd keep his ears open for an Indianapolis production company which might need a new man. Tim, having touched base with his old (and now slightly unfamiliar) friends, drove back to his wife's parents' house where he was to spend the last night of his brief holiday.

The ladies of Ladies' Day have wandered off into the neon-lit vastness of the mall. Ellen talks of that last Sunday. "It was snowing and he said, 'Gee I wish I could stay and play in the snow,' and I said, 'Then stay and play in the snow—work can wait.' But he didn't think that way, you know, act that way. He was going to work Sunday. Walter [King] was picking him up at the airport and they were going to work." In his work Tim drove himself hard, "not over driving but very conscientious . . . too much so.

"He should have been going [home] with me the next day, but he was going back to meet a friend Monday in Washington, and he was supposed to be out of town that whole next week so he felt that he wouldn't get enough work done."

Avalon Estates, Indianapolis. Big houses set back from private roads, landscaped knolls, pruned bushes. It's Sunday. Middle-aged joggers in unfaded sweatsuits, kids on bicycles.

Alice Lindenberg answers the door followed by a small, yapping white poodle. She's steeled herself to talk by dressing as though she were going out. Pale colors to match her eyes and hair. She says she is probably going to cry. On the coffee table, by the couch near the big bay window, surrounded by brochures on the Dalmatian Coast, there is a box of Kleenex. She has discovered, she says, that trips help, and is going to Yugoslavia for eight days in October with her daughter, Lori (the child of Alice and her second husband, Dr. Paul Lindenberg).

Tim . . . your son . . .

"He was probably the most considerate child that anybody ever had, and I think that's the finest thing about him—how he was so considerate of everybody's feelings, so thoughtful. He never went away on a business trip that he didn't call and tell me where he'd be. Now, you can't compare children, but David is kind of on cloud nine, out of sight and out of mind . . . I didn't love Tim any more than David, but there was just a closeness there that . . ."

The poodle barks. In retrospect the dead are seldom anything but perfect.

"They both sort of drifted into the same profession. David was a theater major—though we were really thankful when he got out of theater, strictly theater, because I think it's so tough—while Tim was mainly into radio and television. They began talking of having their own television company some day. David was able to get the company formed—he started out with two or three other men and ended up buying them out—but I was particularly reluctant for the boys to go into business together. They'd always had such a good relationship that, I think very often, when you get into each other's pockets, it can be trouble in a family. And I talked that way with both of them."

She needn't have feared, as it turned out. "They got along beautifully. They had one very good year and then, of course, Watergate just knocked the sock out of them the following one . . ."

Watergate?

"They did a lot of political films, mainly Republican, including a campaign ad for Richard M. Nixon—'To be shown only in the hinterlands,' David once said with a grin—and no one wanted to spend money whether they had it or not. It troubled me terribly when I knew the boys weren't doing well; I used to not be able to sleep and I'd get up and sit on the steps up there with this one [the poodle] who always gets up when I do, and you know, worry about them. Tim knew that I was and knew that I had my nightly prowls, and he'd call to say, 'Mom, we got another contract, you can get off the steps.'" Alice laughs. "So I could sleep for a few weeks till things got lean again.

"He was a quiet person, a private person. I knew that things troubled him, but he didn't say, 'Look, this is bothering me.' He had no patience with his own illness or anyone else's. I'd gone through a period of depression for no reason about a

year or two before [the crash] happened, even enough to
where I sought professional counsel, and Tim was here one
time and I sat down and talked to him about it, and he said,
'I don't want to hear about it.' He said, 'You have always
been the strength in our family and don't tell me you're not
strong.' And so we never talked about it again. He expected
you to get up and go to the party. He had this bad accident
with his leg [one of Tim's shins was fractured in a pick-up
soccer game at college, and since then was held together by
a metal plate] and it troubled him all the time. And he said
to me that day, he said, 'My leg hurts every day but I don't
say anything about it and I keep going, and I expect you to
keep going.' And he said, 'Don't discuss this with David.' And
I never did. You know, when your mother's supposed to be
strong, she's supposed to be strong!"

After her divorce, Alice supported her two sons by working
as an ophthalmologist's nurse for six years until her marriage
to Dr. Lindenberg. "I made a good salary," she says, "but I
didn't have the income I have married to my present husband.
Both boys worked, though Tim didn't work as early as David
because he didn't have to. So Tim was probably about sixteen
before he started selling shoes and things like Christmas trees.
When Lori was born—she was born on the ninth of December—he sold Christmas trees, out in the bitter weather, and
bought her a little red velveteen dress and a little white
blouse. He was big on Christmas and big on gifts."

Lori, now teen-age, blue-jeaned, and blonde, and in the
room with us despite her mother's vague disapproval, says,
"Tim was more like a parent to me than my own father—
mother and father are both elderly." (Slight frown from Mrs.
Lindenberg.) "Tim always straightened me out. I got things
from Tim I really needed . . ."

Thanksgiving weekend was full of parties, long talks, showing off the grandchild, choosing a name for the one about to
be born—Alexis if it was a girl, Timothy if not. Of Sunday,
December 1, she says, "Paul and I were listening to 'Meet

the Press' or something and the program was interrupted by [a bulletin about] the crash of a plane that had been made up in Indianapolis. Well, this was as late as like twelve-thirty, or twenty of one. Of course, I was frightened, but I said, 'Oh Paul, I don't think that's Tim's flight because his flight was at eight-something this morning. Surely he'd be there by now.' Well, I didn't know he had the Columbus layover . . . The kids had stayed here the whole time except that Saturday night before Tim left, 'cause Ellen's father was going to drive Tim to the airport. ["We split it up," Ellen had said, half a holiday with one set of parents, half with the other.] So on the way, Ellen and I sat in the back seat and Tim and Lori and Nicole were up front and Ellen had Tim's ticket out and she said, 'I never will have a joint ticket made up again because we're always having to change them.' I just, out of the corner of my eye, paying, you know, no attention to the ticket, saw 'First Class' and saw 'five fourteen,' which I dismissed, didn't think anything about it. Well, when the news came on, and when they said 'five fourteen' we got in touch with Mike Ahern right away to see if he could get us any information . . .

"We knew. We called over to Ellen's parents and they had heard something but Ellen was asleep. Of course she was seven months pregnant, seven and a half months pregnant. We said, 'We'll come over right away,' and Paul took morphine [along], took his bag with him because we didn't know how . . . what the shock would do to this child. We stayed there quite a while and we just let her sleep until she awakened. And then her daddy told her . . . She turned ashen. She was just in shock. She's been so brave, so brave. She handled this much better than I've handled it.

"You wonder, are all these ninety-two people gathered together for a reason? Sometimes I think . . . I think the flight was mainly made up of young people. Dr. Brown's daughter . . . Tim was in school—they were in the same class, at North Central . . . they were both twenty-nine. Dr. Brown's daughter had just announced her engagement. Another couple had just been married . . . and you wonder. The Bible says our days are numbered . . . Had a flat tire happened, any little thing could have delayed Tim and he wouldn't have been on the flight. So I mean, you wonder about that. Of course you don't see any reason, rhyme or reason for it and you think, it happens to other people. But death . . . Tim's death to me has been so frustrating, I mean, regardless of what kind of life you have, good or bad, you do have some control over it, but you've no control over this. It's like daily I'm banging my head against a wall because I can't see him, I can't bring him back.

"He was just so dear. Just everything about him was dear." Alice Lindenberg cries, and reaching for a Kleenex, scatters the pictures of Hvar, Split, Dubrovnik. "He never forgot a Mother's Day. He never forgot a birthday. He was always there. It's been the hardest thing I've had to do—to give him up. I don't think a mother ever gets over it. It's the last thing you think of at night and it's the first thing that hits you every morning . . . like you just heard it for the first time. I don't know how long that lasts, but it's still lasting . . ." Lori sits by her mother on the couch, touches her shoulder. "I don't have any regrets at all and I'm so grateful that I don't, because we never had any hard feelings or anything I would think of 'If I hadn't' or 'If he hadn't' . . . never anything. It was just pleasant and loving and warm."

On June 27, 1975, the *Indianapolis Star* ran the following column written by Tom Keating:

NEEDED: "BRAIN" WITH A HEART

Living in the age of the computer may be more efficient, but there are times when you wish people had a little more control over the machines.

Last December, Trans World Airlines' Flight 514 took off from Weir Cook Municipal Airport here on a Sunday morning and later crashed against a mountaintop outside Washington, D.C.

Among the 92 persons killed in the crash was an Indianapolis man who left behind a widow and a small child.

After an initial period of intense grief, the young widow tried hard to put back the pieces of her life and not keep dwelling on the tragedy.

This was difficult to do because the crash kept drawing national attention.

Eight lawsuits asking for $9 million were filed in Indianapolis and then refiled in Virginia, where the crash occurred, because that state supposedly has new laws more beneficial to plaintiffs in such cases.

By January, famed attorney F. Lee Bailey was into the act, filing a class action suit in Federal Court at New York on behalf of all the passengers, seeking regular and punitive damages of $850 million.

Bailey named TWA and the Boeing Company, which manufactured the plane that crashed, as defendants in the suit and said he believed pilot error was involved.

Still later, other attorneys claimed Bailey could not represent all the families of the deceased passengers and more legal hassling ensued.

In February, lawsuits totaling $146 million were filed in Virginia in behalf of the relatives of 12 persons who died in the crash.

Big money was involved now, but the young widow didn't file suit, saying only that she would think about it. Mainly she just wanted to forget.

Finally, as the weeks and then the months passed by, the story of the crash drifted from prominence in the public eye as other tragedies arrived with their usual regularity.

But not long ago, when the widow picked up the mail one morning, she noticed an envelope from the airline.

Inside was a bill for the flight on which her husband was killed.

She got in touch with her attorney and he, in turn, contacted TWA and indignantly requested that the billing be canceled and no further notices sent as they had upset everyone greatly.

The attorney was told that a computer unknowingly had sent out the bill and that the machine would be reprogrammed.

Just recently, she received another bill, along with a form letter saying that the money owed was long overdue and prompt attention was needed.

This time the widow's mother-in-law wrote a letter to the airline, trying to get the point across that, given the circumstances, it was highly unlikely the bill ever would be paid.

"There is a computer out there somewhere," she said, "that is causing my daughter-in-law and our entire family great distress by reminding us unexpectedly of my son's death. You'd think someone would be astute enough to correct this, but then maybe that's asking too much these days."

The column is, Mrs. Lindenberg says, essentially accurate. "But can you imagine a company as big as TWA with what is, surely, a tremendous public relations department . . . that the first thing they would think of is, let's pull the charges to this so the families aren't further distressed by being billed. Nobody thought of that."

She wrote Charles Tillinghast, Jr., TWA's chairman of the board.

Did you get an answer?

"Yes . . . I got an answer, from a secretary, on a mailgram. It said they regretted the . . . that that had happened, that they would delete the charges."

"I'd like," she says, "I'd like them to delete my grief and loss."

Two days later, Ellen Harlan and I sit on the floor of the living room in her new house. The house is part of a new development in the northeast quadrant of the Beltway—obviously a rich but not necessarily chic part of town—called Eagle's Nest. We are joined by her two children, the dog, and, somewhat tentatively, the cat.

She says that after the crash all she wanted to do was "to get back to our . . . own thing," to their house in Virginia. "I don't know, I just felt that was where I belonged, surrounded by us and our things." That she wanted to "just get away from all the hassle going on here and, I don't know, I just really . . . Even now I still have a hard time believing it, which is dumb . . . not very realistic . . . That's why a lot of times I go off in my cheery own little world and I just try not to think about it because it's just . . . so . . . like he's on a trip. I mean it's not like he was home every night for dinner. It's like I'm on a little trip and I got home and . . . I was thinking, 'Well if I get back to Virginia everything will be okay, and I'll go to sleep and I'll wake up one morning and it'll all be a dream . . .'

"The first night [back in their house in Virginia] was such chaos. It was like a zoo and I wasn't really ready for that but I kind of blindly . . . you just go through and you're getting dinner and you know they [the house was full of their friends] wanted drinks and I was getting a drink . . . so I really didn't have time to think about it. But I know I kept staring in the mirror, you know, for days and days. I looked terrible, felt terrible. I felt a hundred. I looked in the mirror and wondered if I'd ever look . . . if I'd ever feel that I wasn't a hundred."

Ellen was thirty, a year older than Tim.

"I felt like I'd aged twenty years . . . It was a whole hassle but I was so glad to get back and be with our things. I guess it just gave me some comfort. I don't know . . . just everything, which I still have. When we moved I hung them all back up in the closet because I just can't do anything with them."

Might you marry again?

"I don't really know. I mean . . . I haven't even thought of that. But like some friends . . . they say why don't you get remarried. Well, it's not that simple. It's like, okay, we never argued, I mean once or twice a year you have a [she shouts, illustrating a yelling, door-slamming, I-never-want-to-see-you-again routine] complete bash and everything goes up . . . but we never bickered back and forth . . . I mean it was such a pleasant, such a nice thing. I'm not even *interested* in other men.

"Like if this one— When I just moved in, this one fella who, we've been good, good friends, we'd lived behind each other. It was purely just super good friends for a long time and oh, he'd ask, when are you getting married, and I'd say, oh shut up, you know? Some people say that . . . a lot of other friends, oh, are dating now, or ask you have you met any interesting people—it's the furthest thing from my mind. I just can't even comprehend that. It's just, bahh. I've had a very nice relationship and I think, Gee, if you can have one even once I think you're lucky. We had such a good time and really enjoyed each other and enjoyed Nicole—Tim adored his little girl—at least we, at least the time we did have together was extremely happy."

Nicole chases the dog over the carpet. The baby sits and watches, cheers them on. On the home stretch, they pass the coffee table, knock over a framed photograph of Tim.

Ellen picks it up, puts it back on the table, next to the others. "These were probably the last pictures I ever took of him—we'd just gotten back from Florida. Nicole had a ball and she'd sit in the water and they'd splash around, so I always talk about that . . . I hope if I keep talking about that enough she'll remember him. If I keep repeating— Unless I keep repeating, she won't."

How do you explain death to your children?

Ellen lowers her voice. "You've got to say *something*, you just have to tell them . . . though it took me the longest time to use the word 'died.' " She whispers it. "It took me forever to say that to her. It's just, well, Daddy isn't here. For a while that was good for a stall. I told her that well, [people who died] just aren't here anymore like we are. One thing of course is that she doesn't know 'how.' There's no reason for her to. So I try . . . not to say anything about that."

Some of the neighborhood's older children were told of the crash by their parents and, Ellen says, "They've asked me about it and we've discussed it. One of them said, 'I know what happened,' and I said, 'Yes, you may, but Nicole doesn't, so don't you mention it.'

"The big question though, is '*Where* is dead?' But where? What do you do? Where do you go? I told her I don't know, I don't know! That's what I have to tell her because I certainly can't give her any, you know, religious thing because I'm not sure *I* believe in all of that.

"And there aren't very many books written on this. I went to the library because I . . . because I want to tell her, I needed help to help her. There are a lot of books that tell you what *not* to do, but that still left a lot of unanswered questions."

Ellen says she has talked over the problem with Alice, and with her own parents, "but nobody has any, you know, marvelous answers . . . In some of the books I did read, they said that a child can get mad at the mother, or the surviving parent, like they did something that caused it. Or just mad at the fact that everybody else around is saying, 'I have a daddy, where's yours?' You just have to know your own child a bit and know what they can handle and maybe use something in their realm of understanding.

"Another problem that bothers me is that, obviously there aren't any pictures of the baby with Tim." Ellen says it evenly, yet another legacy of death to be coped with. She says she's told Nicole, "When Alexis grows up you can tell her all about your daddy, and it will help her know your daddy, you can tell her all about him." Nicole, she says, "thinks that's great."

On one of her rounds Nicole has taken some of Tim's pictures and, right now, as though on cue, hands them to Alexis, who, with her small pudgy baby hands, grabs the frames, starts to put them to her mouth.

Ellen, gently: "Honey, we can't let the baby have the pictures—she doesn't know what to do with the pictures yet. She's not a big girl like you."

ROBERT L. MURDAY, 48, BODY BAG #101

"This friend of mine was saying that since Bob and I hadn't had too much time to spend together and that there were so many things we didn't get to do—things we'd planned on doing—that's what is making it harder to get used to." Jean Murday lights a cigarette, one of many, shrugs—narrow shoulders hunching under synthetic wool—offers her own appraisal of loss. "Of course, life's not as nice as it was before. I . . . I've always been used to having something to look forward to and now, I really don't. If you know what I mean. But," and she's been wanting to say this all along, "you have to face it. Everybody's going to go someday."

But when it happened, it was in fact so brutal that, instead of forgetting or blurring its particulars, she remembers them with icy, almost detached, clarity. In a single morning, between six and noon, the two men with whom she had shared her life died: her first husband cancer-ridden in a hospital bed, her second fractured among the burning wreckage of a jetliner. December 1, 1974, was not, she says, "a normal day any way you look at it. It was not . . . a normal day."

Backtrack: After two years of hesitant courtship, Jean and Bob Murday were married in Columbus, Ohio, on December 23, 1972. She was forty-nine, worked in a supermarket, had been divorced for seven years, and lived with her grown son in a modest house off East Broad Street, out where the old estates and private schools end and the shopping centers begin. Murday was forty-six, had never married, and lived at home with his elderly parents when he wasn't on the road installing commercial baking ovens. When he moved into the house on the corner of Seigman Avenue, Jean promptly told him that her first husband had "always hated it—I wanted him to come in here and feel at home"—told him that she understood and accepted his being away three weeks out of every four, and just hoped he'd call.

No, he was not a traveling salesman, she says, he was a "consulting engineer," a man with a title. "Every year he got a thousand-dollar raise, at least a thousand dollar raise," and, "at the end, was making sixteen-something."

They'd met at a neighborhood bar called The Silent Woman, where "a group of us all got acquainted, played the bowling machine" and such. "One night I turned to him and said, 'How come you never asked me for a date?' So then he brought me home and the next day I was so embarrassed and thought, 'Oh, he must think I'm terrible.' But the next night he called me and came over and that was it." A tallish, perfectly pleasant-looking man with the slightest hint of a paunch, large black-framed glasses accentuating the length of his vaguely stern face, Murday was, she says, a man to whom "I wasn't married long enough to ever fight or anything . . . I had, you know, all the good things of a marriage."

We sit in the kitchen of the house on Seigman Avenue drinking instant coffee and smoking cigarettes. The ash tray betrays her apparent calm and sensible acceptance of irrevocable fact.

The last weekend?

Friday—"Bob got me my dishwasher for Christmas last year. We had gone over and picked it out, let's see, on Friday night, and he gave me a check for it and they were to deliver it the next week, on Wednesday, and he was going to install it when he came home. He had all the pipe and everything. He had it all planned, how he was going to fix this up for me and that was my Christmas present. I had told him I really didn't know what I wanted . . . He just wanted me to have a dishwasher, he was always wanting me to have things. I never really minded washing dishes but . . . I says, 'Well, there isn't anything I really want for Christmas.' I said, 'Why don't you just go ahead and get the dishwasher,' and I said, 'then you can get me a little something personal if you want to.' So then we went over and picked it out. It was delivered the next Wednesday after he died so I had to call a plumber to have it installed . . . That plumber, this couple that Bob and I used to run around with—we played cards a lot, we'd have more fun at card games . . . 'cause everybody played so crazy. So this man came and put it in while I was at work and he never did send me a bill on that, and I thought that was so nice of him. It really was."

Saturday—"Bob had been out and picked up his dry cleaning and I had worked till three and came home and was starting dinner when he came in . . . I can still see him come in the back door there, and oh, we were kidding about something and I says, 'Oh, out having a beer again?' which I would kid him about, and he says, 'What you gonna have for supper?' and I said, 'Why don't you fix some of the shrimp—' "

(Jean had been trying to teach him how to cook because "I thought what if . . . you know, I'm older than him . . . What if something happened to me and he needs to know how to take care of himself? He'd always been at home with his mother to take care of him.")

"—'cause I had bought a big package of the big shrimp, you know, frozen. And I says, 'Well, if you wash them we'll have them 'cause there isn't too much to cooking them. You just put them in cold water and when they come to a boil, take them out.'

"We had shrimp salad for our dinner," she says and Bob went to pack for the next day's trip. "He had it down to a science, packing, and he took his big suitcase because he was going to North Carolina and cigarettes are so much cheaper there, he was going to bring some home. He had a lot of little hints for traveling . . . He said he'd been in some hotels that were really cold in the wintertime and he'd run out of underwear and socks, and he'd wash them out and he knew more ways of drying them, hang them on the light, or on the television, he'd say, 'It's surprising how much heat you can get out of a television if you turn it on.' " She laughs. Memories of small things.

The Murdays then went to The Coachmen for a drink. It was a friend's twenty-fifth wedding anniversary. "We sat there and talked and I said, 'We might make it to our twenty-fifth, but we'll never make it to our fiftieth.' And I know this one girl, she said afterward, it made her feel so bad because we had been talking about it that night."

Sunday—The phone rang. It was six o'clock on a gray, snowy morning. Jean Murday rolled over and picked it up. Her daughter, Patty, was on the other end. "Mom," she said, "Vicki just called and told me that Dad died."

Jean lay in bed, swallowing, trying to blink away the twenty-two years of marriage jumbling through her mind— young womanhood, childbirth, encroaching middle-age, frustration, the ugliness and suspicion at the end ... ("Even if you're divorced from somebody, still ... if they die ...")— then rose and slipped downstairs.

Murday dressed for the trip, blue shirt, striped tie, gray slacks, blue blazer, and followed her. "Have you told Bobby?"

"No," said Jean, "not yet. I thought I'd wait till he woke up."

"It's still his father after all. You should tell him."

Jean woke her son.

"I called the hospital at three," Bobby answered. "They said he was all right ..."

"I know," she said. "I mean ..."

Patty arrived with her own two children. Brief tears. Leaving her boys with their grandmother, Patty drove to the hospital in the welcome privacy of her car. Jean fixed breakfast. "Then," she says, "we sat."

Finally, it was time for Jean and Bob to drive to the airport. "Look at the snow. I wonder if the plane's gonna take off." Murday parked in the slush behind the terminal. Snowplows, yellow lights kaleidoscoping, belched exhaust into the gloom. "Oh, that won't hurt anything," replied Bob. He tugged at his bag, got out. "The only thing they worry about is the snow on the plane." Jean slid over into the driver's seat. "You're gonna call on Tuesday night aren't you?" She drove back concentrating on the road.

"So then I got home and, I don't know, the more I thought about it, the more I thought I should have said something. I feel I kind of neglected him that morning. I was feeling bad about my first husband, I had my grandchildren to watch and it just seemed . . . I didn't pay too much attention to him. I thought, 'I really should have said something,' because though I felt bad about it I didn't want to show it, see because . . . He might have had more understanding if he'd been married before, but this was his first marriage . . . you know, I didn't want to hurt his feelings. I just had mixed emotions, but then I thought he's gonna think I'm real hard-hearted if I don't show any feelings at all, and it just went through my mind what I was gonna tell him when he called Tuesday night. I was going to tell him, 'I don't want you thinking I'm so hard-hearted, I did feel bad about it, but I didn't want to hurt your feelings either . . .' I just couldn't wait for him to call me."

Seeking refuge from her thoughts that afternoon, Jean took a shovel and cleared the driveway. Chilled, she then went upstairs to make the bed. The phone rang—it was Patty again. "Mom, where was Bob going this morning? Was he going to Washington?"

"Yeah," Jean said, "he had to change planes there. Why?"

Patty said her husband heard on the news that a plane from Columbus had crashed near the capital.

Jean asked "What time?"

"Ten-something," Patty said.

"That couldn't have been his—his was supposed to leave at nine thirty-five. But I'll turn the radio on and listen; I'll try to find out."

"Yeah."

"I'll try," Jean said again. Then, softly, "but nothing else can happen today . . ."

The early radio bulletins were inconclusive.

Patty called back.

"If I call, if I make any calls I'm gonna cry," Jean told her. "Nobody's gonna make any sense out of it. You call."

TWA was not helpful.

"So then we waited till six o'clock," Jean says. "That's the time he'd be in his hotel. We called where he was supposed to stay and he had never checked in."

The phone rang once more.

"Is this Mrs. Robert Murday?"

Jean couldn't place the voice.

"This is the Columbus *Dispatch*."

"Yes?"

"Is this the Robert Murday family, the one that was in the airplane crash?"

"Yes."

DR. ALBERT GOLDBLATT, 60, BODY BAGS #21 AND #6-A
ADELE GOLDBLATT, 59, BODY BAGS #41 AND #223

When *The Washington Post* ran the passenger list and the attendant column of obituaries, the Goldblatts, as befitted home town residents, rated the most space. There were also two photographs, no doubt taken at some forgotten ball or charity gala. One showed a smiling man with a small mustache and dark eyebrows; the other, a meticulously coiffed blond woman. He is wearing a dinner jacket and a bow tie which does not look like a clip-on; she is chic but not ostentatious. Here is their obituary:

Dr. Albert S. Goldblatt, a Washington dentist for 37 years, and his wife, Adele S. Goldblatt, of 3308 Glenmoor Drive, Chevy Chase, had gone to Indianapolis on Thursday morning to spend Thanksgiving with a son, Dr. Lawrence Goldblatt, an oral pathologist, his wife and their two children.

Both Dr. Goldblatt and Mrs. Goldblatt were natives of Washington. He attended old Central High School and was a graduate of Georgetown University and its dental school. He had offices at 4500 Connecticut Avenue NW.

Active for many years in the D.C. Dental Society, Dr. Goldblatt had served on its executive committee for the past 10 years.

He was also a member of the Board of the Society's Research and Educational Foundation and had been elected chairman of the annual convention scheduled in 1976. He belonged to the American Dental Association.

Dr. Goldblatt was a charter member and past president of the Maimonides Society, a dental group that provides interest-free loans to dental students at Howard and Georgetown Universities.

He was also secretary and a member of the board of the Dental Health Services Corp. of D.C.

Mrs. Goldblatt grew up in Weehawken, N.J., but returned here to attend George Washington University. The Goldblatts were married in 1937.

She was active in the Women's Auxiliary of the D.C. Dental Society, was a past president of the Maimonides Dental Auxiliary and had been a volunteer worker at the National Children's Center and the Red Cross.

Both had been active in the Selby Bay Yacht Club at Mayo, Md., where Dr. Goldblatt, an avid fisherman, was a charter member and officer. He had been active for many years in the U.S. Coast Guard Auxiliary.

In addition to their son, they are survived by another son, Dr. Norman Goldblatt, a physicist of Rochester, N.Y., and two other grandchildren.

Dr. Goldblatt is also survived by his parents, Mr. and Mrs. Hyman Goldblatt, and a brother Harold, all of Silver Spring.

Mrs. Goldblatt also leaves her mother, Jean Sisco, of the

home in Chevy Chase, and a brother, Bernard Sisco, of Kensington.

Everyone else walks in sweating—it's mid-August in Indianapolis and the businessmen who lunch at Stouffers Inn order cold beer and gin and tonics. Larry Goldblatt has hot coffee and a chef salad.

He's in his thirties, bearded, neat. As disciplined in the emotions he allows to surface as he is in his eating habits, he quietly chronicles his family's progression up the professional and economic ladder, beginning with his paternal grandfather, a Jewish immigrant from England who worked as a salesman for a Washington food distributor, to his own parents, hardworking, self-made people ("You hear that about everybody but my father did work his way through college and dental school, worked in a drugstore and, as I get the story, he worked very often approaching twenty-four hours a day between school and his job"), who had struggled to give him and his brother—a physicist—the proverbial advantages they'd never had. They'd done it discreetly, laying no burden of guilt or enforced gratitude on their children, just as they had cared for and attended to their own parents, who were, Goldblatt says, shattered by the crash—elderly people whose link to the present was brutally destroyed. "My father was the figure to whom everyone looked in times of happiness or times of sadness, times of need as well as times of sharing good things. Even though my father's parents lived in their own apartment, my parents' presence was always felt; it lent them a great deal of security. If anything went wrong, they knew they'd always be there and stuff . . . Of course my grandmother lived with them, she was almost like a third parent, so she was very hard hit by this. She lost her home."

He considers his own sorrow. "My wife and I have a very good marriage and I think we're very close to each other—she's the one with whom I'm closest but there's . . . it's very lonely, you know. I think that—not to lessen anybody else's loss—but when you lose your parents when they're old and they've lived a full life, when the pendulum of giving emotionally and otherwise has swung from parents to children to the other direction, when that happens, it's very sad to lose them, very, very regrettable, but it's not truly tragic because it's a natural thing. But when parents are lost at the peak of their potential to give emotionally of their experience, and also are at the peak to receive from children who are just beginning to be able to give back, that's tragic.

"I suppose we should be thankful to someone or something that the past was, on the whole, very happy. But it's what *isn't* that's the problem: the opportunities they didn't have, the fact that they worked so hard all their lives. My father was a dentist, he made a good living, but he wasn't particularly fortunate in any meager investments he might have made; seemed like everything he invested in immediately went down. I asked him, oh, must have been last year, when he was going to retire, and he would just laugh and say, 'I'm not gonna retire, not in the near future.' He said, 'I couldn't afford to retire,' and we'd kind of laugh about it and I wouldn't know whether to take him seriously or not. I never questioned him in detail about his finances, but I am the executor of both estates and, my God, he was telling me the truth. What he made he spent, keeping everything together. They did . . . they did quite a bit of traveling; they generally took at least one trip a year. They went to Europe a couple of times, or they'd go down to the, I guess the Caribbean Islands, you know, for a week or something with good friends, but there wasn't much left over, and they did that mostly in the last few years because those were the years when they weren't saddled with the expense of sending a couple of kids through school. They sent my brother through school, through college, and they sent me through college and dental school. I tried not to take advantage of it but whenever I needed money, it was always there. And I didn't know from this working twenty-four hours a day in the drugstore business because my father would never . . . but it took everything he had to do it.

"I was sad for my father and mother that they couldn't retire any time they wanted to by a long shot—" the *they* is because Mrs. Goldblatt had recently taken to helping keep up the bookkeeping end of her husband's "one-horse" practice as Larry called it—"not that he would want to retire, really. He wasn't ready to quit work. He loved his practice. He was a fantastic dentist—he was very good. He was my dentist." A gentle laugh. Goldblatt has a mouthful of perfect teeth. "It has a lot to do with genes."

Talking of which, he says that he feels deprived of "at least twenty-five years' worth" of his parents' company. "If you could judge longevity from parents to children, why, I'm sure my father would have lived a very long time. My mother perhaps would have . . . I don't know—her father died young but her mother's still certainly very active . . . There are so many things that you want to tell them, things you want to do with them and for them . . . But, I think that the loss that my brother and I have suffered, as heavy as it is, doesn't compare with the loss that my grandparents have suffered."

This is the insistent note in our conversation, an unselfish and concerned awareness of the emptiness injected into the lives of the older members of the clan.

"I've thought about this a lot. Some people lose their children in miscarriages, some children are born with congenital defects, some are lost early in life by accidents, by violence . . . and it's difficult to gauge who suffers most. It's pointless as a matter of fact. But there was something about seeing your children, I would imagine that there is something about seeing your children rise to a point where they are, you know, where they have the most to give and are at a great point in their lives which is not only filled with energy to give of themselves but also with a great deal of experience to give to others—"

This is what the statutes mean when they allow damages

to be recovered for loss of solace, comfort, and advice. Those are not just legal terms in some code of law.

"—to see them at that point, to be cheated out of the rest of their lives, you begin to wonder—I've begun to do a lot more than wonder, you know—what possible purpose something like this could serve. What good could a plane crash do—in somebody's mystical scheme of things or ground plan or whatever it is? I don't see any. So I think the grief of losing children such as this is the worst part of the whole thing and I know that I can't even begin to comprehend their grief."

Lunch is over. We walk outside into the sun broiling the flat, shimmering city.

Are you angry?

Yes, he says, very. So much so that he won't even deal directly with TWA or its insurance company. "But the worst part of the anger is its futility. There's nothing I could have done to prevent the crash, and there's nothing I can do now to remedy it."

NANITA FAYE MATHIEU, 26, BODY BAG #104

Nan Mathieu's husband, Mario, is a Chilean émigré with a bushy red beard, blue eyes, and an unmistakably European cast to his South American face. He came up north, along with his family, at age thirteen. Educated around Washington, D.C., he wanted to be a doctor but dropped out. He has a soft, Latin way with English and talkative hands. Drafted into the army in 1971, he was posted to Fort Campbell, Kentucky.

"That's where I met my wife, Nan. I was the company clerk because I was the only guy who could type. She was a recreational therapist at the Fort, working at the enlisted men's service club. I got there in February of 'seventy-three;

she'd been there since December. I'd volunteered for airborne training in Louisiana to stay stateside and applied for a transfer to Fort Bragg after that to be nearer to home. But then I met Nan and went back to Kentucky after jump school.

"She was born in Logan, Kentucky and had gone to Eastern Kentucky University, majoring in speech and recreation. When she got out and looked around for a job, the army was hiring civilians, and she took the job. She thought the army, well, you could travel, but no, she took the big step instead and married me . . . Before that she used to invite me over to the house, we were good friends, nothing more. I used to say to myself, 'It's a waste of time, why compete with twenty thousand other guys, what's the sense?' So she used to invite me over and I never went. If you're not going to do anything, why go?" Big smile through the red beard.

"She really got mad at me one time. She said she had to go to Nashville, to play tour guide for a group of soldiers, and that she needed one more person for the bus. She told me that if I didn't go, she wouldn't talk to me anymore. So I said to myself, 'Nashville? What can happen in Nashville with fifty other guys?' But I went and we sat together on the bus and started talking. I said, 'I don't want to get married till I'm around forty.' I just said that and she agreed, she said that she was a career woman more than anything else, so fine . . . we got along. We talked all the way there, we were together all day, and on the way back we talked some more, and she's the one that says, 'Are you a good dancer?' And I said, 'I do my best,' and she said, 'How would you like to go dancing? As friends. Tonight, in Nashville, we'll go back to Nashville.'

"It's fifty miles away, it was a Sunday, so we went back. She had a car—she drove me around for six or seven months till I got my car—I had a little money, so we went dancing. It was the first time I'd seen her in civilian clothes, believe it or not. Before that I'd always seen her in her civilian service uniform, you know, the one with the light blue skirt and white blouse. It didn't look that bad, but she had hair down to her waist and I didn't know that. And we went dancing. Then I liked her and I got fresh that first night and I got smacked. I figured why not? and I got smacked and then she was mad and wouldn't talk to me, thinking I guess that the proof was in the pudding, that you can't ask one of these soldiers out and expect them to behave like civilized human beings.

"We'd gotten back around two A.M.... She didn't have to go to work till four that afternoon, and I decided if I don't make a pass now, I'll never make a pass so... I made a pass and she didn't like it and smacked my face. Thwat. I liked it I guess, I think, because I went back to her. It took me a while to get her to go out with me again. I had to tell her all I wanted to do was apologize, so I took her out to dinner. I had to borrow a coat and shirt; I had pants but no coat. And a tie too. I really behaved like a gentleman then, I really did. To show her that I wasn't only after her body. After that I said, 'I wish I could see you again.' All I wanted to do was to get back at her, you know, to try and make her go out with me, try to make her fall in love and then say 'Goodbye, I'm leaving.' I was hard core. But it turned out okay, I don't understand how." Mario's eyes soften as he tells of that first night's fracas.

"When I met her, I was a PFC and she could have had her pick of any officer, anybody, and she picked me, I guess, a nobody... Of course it was nice..." His voice rises and falls as he remembers, a hint of laughter at the ends of words, sentences.

After a year, they married, Mario defiantly out of uniform ("no soldier suit for me"), and moved off base. "It was a hassle in the beginning. I had a couple of fights. I didn't have any fights when I was single. But after we were married, a couple, because of where she worked. Not that I was jealous; I'd do that for any woman. I remember one time, in the pool room, someone said, 'Gosh she's got a big ass,' so I'm not going to hit him right off the bat. I said, 'Look, that's my wife you're talking about. I think you'd better take it back,' or something like that, and he got wise and said, 'Well, I still think she's got a big ass.' To me, this isn't gentlemanly-like, so I just hit him and then I had to explain to my wife's boss why I did it.

"But you know, the infantry is full of Puerto Ricans and blacks, and the ignorant whites, and the hillbillies, the rednecks—that's what it's made of, the infantry. But they're easy to get along with because they're narrow-minded, you don't have to do much to please them. But they're good people, and usually they take care of somebody else's wife when he isn't around."

Discharged in 1973, Mario came home to the Virginia suburbs of Washington. He took a job with TWA at National Airport as a baggage handler because he needed the money. Life was expensive: a new marriage, the transition from a military to a civilian regime. Nan applied for a therapist's job at St. Elizabeth's Hospital, the District of Columbia's public mental institution—a long row of walled-in, red brick buildings on a hill in Anacostia, in which, for twelve years after World War II, the poet Ezra Pound was confined.

"When they first offered her the job, I thought, 'It's too tough.' I'm from this area. I said, 'No way.' I said, 'I don't mind, you can be a waitress somewhere, you can forget your degree, forget whatever you went to school for, because you're not going to work there.' But she convinced me to go there and see. I did. I went with her when she went for the interview, and I saw the people that work there, and they were all young and smart . . . mostly girls, believe it or not. It was nice, the people were nice, and she liked it very much and wanted to do it. That was her life, work like that. So I felt stupid and said okay.

"It was funny sometimes . . . We used to talk about it. I guess like any other human being, any time you talk about the mental . . . about mental retardation, you know, the things they do, I think they're funny. I don't mean bad or anything, but sure we used to laugh. And in the summer she'd spend a week in Maryland at Camp Happyland, a summer camp for the patients, eat with them, sleep with them . . . I went too. Wherever she worked I did volunteer work. That's where it started, at St. Elizabeths. They needed some drawings, like for a story in their newspaper about a sick tree, and I would make a picture of the sick tree. Then, when they needed something bigger, they'd call me.

"I used to work nights at the airport, four to twelve, and she worked days, nine to four-thirty, so by the time I was going to work she was coming back home. It was like this until six months before the accident. We . . . we weren't getting along as we should—we were getting on each other's nerves because we weren't seeing each other. We used to have fights all the time, little things. I was really fed up with being married. I wasn't seeing her anyway, so what the hell was the point? So I changed my hours, to work days to show her we couldn't get along anyway and . . . what happens? We get along great. Because we were sleeping together . . . for a change, eating together. 'Wait until you get to bed and everything will be forgotten—you train them that way.' No, that's chauvinistic [laugh], but I've heard that from my father. He says, 'In a happy marriage you can fight all you want during the day, but when it comes time to go to bed, you kiss and make up,' which is true. You can't go to bed with a person lying over there mad.

"It worked out great, because for the first time since we were married [nearly two years by then], we worked the same hours. Before that it was always night shift, night shift, night shift. Then we were the most partyin' couple . . . We didn't like to stay home. We'd do all kinds of things, even go to the playground and ride on the swings. We had fun, were really like kids, going out in the rain, the snow. People probably thought we were silly."

The laughter fades. Mario talks about the future they'd hoped to share—Nan staying at St. Elizabeths for a while, then having a couple of kids and, when they were grown, going back to work. The government perhaps, a civil service job. A long pause . . .

"This is a story I don't tell anymore... Well, anyway, Nan had an operation, on the ovaries, a cyst, she had it removed, and they told her—this is what I found out later—they told her there was a fifty-fifty chance she might not have kids. I didn't realize it then. My sister told it to me after everything was over. My wife never told me because she knew I wanted kids. Well, after the operation, we decided it was time. We talked. We could afford it, we were both making good money, we didn't have any big debts, so we could afford a kid, so we said, 'I think it's time.'

"So she wasn't taking care of herself. If it happened, it happened; if it didn't, fine. And then I started seeing her getting sick in the morning, and I didn't pay any attention. Me, I'm so absent-minded... Now I go back and think of all the things she was doing, like they say in the books, eating all these crazy things—she'd want to go out and eat ice cream all the time, or pizza. One time we were at a party at my uncle's house and I had to go to work—I was working one of the presidential flights. I went over to Andrews Air Force Base—every time the President flies, there's a plane for the press, usually TWA, so we are supposed to go out there and load and unload, which to us is money. Overtime or time and a half. So I went out there at midnight to take care of the press flight."

Mario got home around 3:30 in the morning. Nan woke to tell him she'd passed out at the party, "just felt dizzy and fell on the floor." It was then, Mario says, "that I started thinking, she might be, you know, pregnant."

Six weeks later Nan left to spend Thanksgiving in Lancaster, Ohio, with the aunt and uncle who had raised her while her parents went through their years of predivorce squabbling. Unable to rearrange his work schedule, Mario had had to stay behind.

"When I called her Saturday night, we laughed—we used to goof all the time on the phone, make silly voices and all that—and she said, 'I wasn't going to tell you, but, uh, I went to the doctor here and I'm pregnant.' That's what she said to me. I said, 'Oh great,' and everything and told all my family.

"So now you tell that to people, like TWA, and they don't believe you, they think you're saying it for the lawsuits. But to me, she was. Because I'd heard it from her. She'd gone to the doctor, I don't know which one, I don't care to know. You have it in your head that she did, that she was."

Up at the cold makeshift morgue in Bluemont, there was no uncertainty. The pathologists had found a two-month-old fetus along with the rest of the contents of Body Bag #104. Another little horror added to a lengthening list.

"It's lucky for me that I never really felt like I was a father because I never got to see her. If I would have, I'd have felt terrible, more so than I felt, which is bad enough. I didn't really realize I was going to be a father."

Later, much later, Mario and his friend Carlos Gomez realized, but did not discuss, that Gomez (also a TWA baggage handler) might have carried that body bag. Gomez had volunteered to go up to the mountain along with the other TWA crews. He felt he had to. Mario was his friend, Mario's sister his bride-to-be.

Carlos, a Cuban refugee who spent his first three years in America in a Catholic orphanage in Peoria, remembers driving up to the crash site at 6:30 Monday morning. "I just got out of my car and there was reality ... everything ... pieces ... you name it ... They gave me a plastic bag, a garbage bag, and gave me surgical gloves, and told me to go around with a stick—there were all kinds of leaves and rocks and snow— and to look around, and if I saw pieces, to put them in the bag."

Gomez had brought his movie camera along. He asked the FBI agents in charge if he could shoot. Busy with other things, they said okay. "I said to myself, this is something ... out of the ordinary. But I made a point to myself just to film when everything was picked up ... it's something to keep. I was there."

Detailed to the morgue, he says, "I got there and started shaking again. I thought I'd seen everything at the site. Thank God it was winter time, and that all the windows were open. They'd set up some black vinyl covering on the floor . . . There was at least an inch of blood on it." Thinking of his flight from Cuba at the age of fifteen, his years in the Peoria orphanage and those in the Spanish ghetto of New York, Gomez says that until that moment, he believed he had already "been to hell and back in this country."

Mathieu stayed off the job for two weeks after the crash, then went back, but "my friends didn't know how to talk to me. Nobody would make a joke and that's what I needed, to be the way we were . . . I didn't need the silent treatment. So I was really a loner for a long time until they saw me laugh, and then they started realizing I was normal.

"There were some people there who were extraordinarily nice to me, but my direct bosses, they . . . there are no words to describe, to say . . . If I was there I was supposed to do the work as though nothing had happened. I couldn't give a damn, really. I went back because I had to."

But having found a job with Lan Chile Airlines, Mathieu quit his $12,000-a-year TWA job a month later.

"Somebody once told me that you can never forget somebody you lost, but that you have to learn how to live with it. I guess that's what I'm doing now, learning how to live without my wife. The way we figure, you get married once and that's it. I'm not saying I'm not going to get married again. That's ridiculous, because I love kids and I want to have kids of my own and I have to get used to the idea that I'll meet somebody else, that there'll be another woman in my life. I know that, but up to now [nearly a year after the crash] there hasn't been anybody at all—you think I'm kidding . . . I got two friends that I go out with. One is my sister's best friend, a good friend of my wife's, and the other is Carlos, my brother-in-law. I don't want to go out on a one-to-one basis. I used to love to with my wife, or before we got married.

I thought it was a challenge, like a bullfighter and a bull, but now I don't, because it's too ... it's too compromising. You feel like you have to *do* something."

Mario Mathieu tried to satisfy his need to do something by buying a new house in a suburban Virginia development and a bright orange Volvo sedan, and by quitting his job with Lan Chile. Then for two months he sold gravesites—five of them, he says, on a 15-per-cent commission basis—at the cemetery where he had buried his wife. No, he says, he never used that in his sales pitch.

Mathieu had planned to take his wrongful-death suit against TWA all the way to trial, having scorned their original settlement offer of $150,000 as "ridiculous." Spooked by the thoroughness of Associated Aviation Underwriters' investigation into his private life, he settled just days before the scheduled trial.

"I'm not after the money," he says, though "the money is going to help me, of course. But if I were to die, I'd want my wife to have things ... I'm not after the money, but TWA treated me so bad."

He had nightmares, a recurring dream about his wife calling him on the phone. "The phone would ring, I'd pick it up. 'Matz,' she'd say—she never called me Mario, always Matz, I thought it sounded like a dog but—'Matz,' she'd say, 'don't be scared, but it's Nan.' 'Where are you?' I'd ask. 'All I can tell you now is that I'm not dead. But I can't come home yet, but you'll see me soon.' And that would be it and I would wake up. I went to a shrink and he said it might be because I talk to Nan's aunt and uncle a lot, that every time I talked to them it reinforced the thought that she wasn't dead. He told me that I should go and see them. I didn't want to go through that again, but I went out there and really had a good time. It was like a normal thing—we ate, went shopping, got out Nan's baby pictures, and, believe it or not, we laughed all the time, at all the silly things she did when she was a kid. After that I never had the dream again."

Mario's face clouds as he remembers another nightmare—the funeral and the problems leading up to it. Nan's mother, divorced from her husband for a number of years, apparently insisted that her daughter not be buried in Virginia. Asserting himself, Mario told her, "I'll bury her where I want to. If I want to bury her in my back yard, I'll bury her in my back yard and you won't have anything to say. I'll do it my way, the way I want to.' The thing is, I was going to have her . . ." he searches for the word, ". . . cremated, because after a wreck, what can you do? I figured it was the best thing, the cheapest thing, 'cause we're talking about *money* now . . . Not that I'm cheap but I didn't have any money then . . . But her aunt and uncle were firmly against it. They wanted her buried. If the mother had said, 'Don't get her cremated,' I'd have done it just to go against her.

"So, I needed two thousand dollars to bury my wife and I went to the TWA credit union. They wouldn't give me the two thousand—they'd only give me sixteen hundred because I owed four hundred on another loan. So I had to sign a letter saying I'd put my wife's insurance up as collateral before they'd lend me the money. See, that's what I mean about them.

"It's a weird thing . . . You don't borrow money from your family to bury your wife."

TRANS WORLD AIRLINES
605 Third Avenue, New York, NY 10016 U.S.A.

Vice President

Flying

TO ALL FLIGHT CREW MEMBERS

Once each year, Flight Operations presents a special issue of Flite Facts dedicated to emphasizing safety and safety awareness.

As we move through October, traditionally TWA's safest month into November, among our worst, the industry has lost nine jet aircraft thus far in 1974. This includes two Pan Am 707s; two TWA 707s and most recently an Eastern DC-9 at Charlotte. Except for our own performance this represents a considerable improvement over the same period last year when 16 aircraft were lost, including one Pam Am 707 and one Delta DC-9.

Of significant importance is the fact that in 1973, 86 percent of the 29 hull losses occurred during the approach or landing phase of flight just as they have during almost every year since the introduction of the jets.

Although not a factor in our recent loss [the 707 crash in the Ionean Sea], a very sad fact concerning accidents of this type is that the majority involved the crew through either poor planning, lack of proper cockpit discipline or deviation from well-established company policy and thus were avoidable.

An in-depth critique of our accidents indicates to us that we have had adequate altitude awareness to minimums and excellent crew coordination procedures. However, we need to emphasize the "Approach Envelope" and increase alertness to height above the terrain in relation to the horizontal distance from touchdown. To help re-emphasize the need for safety awareness, a TWA/ALPA [Airline Pilots Association] Safety Committee has been established to provide the base for continuing cooperation between management and ALPA. Among the accomplishments of this Committee has been the establishment of revised crew coordination procedures. These revised crew coordination procedures are designed to relate the 100 foot altitude point to a fixed horizontal distance from the touchdown zone. Through repetition, the use of visual cues at this altitude and the associated call-out "100 feet" will aid in maintaining a stabilized platform on that portion of final beyond decision height. This, it is hoped, will eliminate the problem we have encountered of landing short.

Safety is the very foundation of all our practices and upon which our future and that of TWA depends. To properly insure this future, we must develop within our cockpits an environment of caution and conservatism.

Sincerely,
(signed)   R. L. Simpkins

—PAGE ONE OF THE FALL 1974 ISSUE OF FLITE FACTS, TWA'S IN-HOUSE LOG, PUBLISHED AND CIRCULATED ONE MONTH BEFORE THE CRASH ON THE BLUE RIDGE.

Trans World Airlines might have been more generous with Mario Mathieu's request for a loan with which to bury his wife, but the Company was in desperate financial trouble. While the rest of the giant TWA conglomerate (Hilton International Hotels and the Canteen Corporation) made money, the airline itself was to declare a year-end, before taxes loss of $45.9 million—the equivalent of a brand new Boeing 747 Jumbo Jet and a brace of 727s two years' worth of unemployment checks for all the jobless in the state of Indiana. Only TWA's domestic flights were in the black— by a paltry $183,000—yet down $20 million from the previous year.

Bucking huge rises in the price of jet fuel, the legacy of a crippling flight attendants' strike, and a general, recession-induced slump in air travel, TWA was, in December 1974, an airline fighting for survival. Its 3600 pilots (many of them, including Flight Engineer Thomas Safranek, forced onto reserve rather than regular line status) lived in fear of layoffs, its stockholders once again failed to receive quarterly dividend checks from their sinking common stock, and F. C. Wiser, its president, was about to be out of a job.

Cutbacks and economy were the corporate passwords. Crew schedules were shuffled and flight dispatchers were intensely conscious of additional expenses incurred by rerouting planes.

Alan Clammer, TWA Flight Engineer and Tom Safranek's friend: "We have some management personnel on flight operations at TWA that I think are less than totally safety-conscious. For example: the crew of Flight five fourteen—this was a 727 and we do not operate 727s into Dulles Airport. 727 crews know Washington National very well. They don't know Dulles. So here's a crew on its way to Washington National and the phone rings [in the cockpit] and it's the company calling, telling them that Washington is closed. In reality, they should have had the option of going either to Baltimore or to Dulles. This was briefly discussed with Mr. Epp the dispatcher. Mr. Epp opted for them to go to Dulles. It was easier to move the passengers—cheaper—we had better facilities at Dulles from a passenger standpoint. What this crew did not know was that the Primary Navigational Aid on runway twelve—namely the Armel VOR—had been out of service earlier in the day. A NOTAM [Notice to Airmen] was issued stating this. The crew of Flight five fourteen never received this NOTAM. Secondly, there was a SIGMET [Significant Meteorological Advisory] issued calling for moderate to severe turbulence, icing, clouds, extremely heavy surface winds. The kind of weather that pilots do not knowingly fly into. The TWA meteorologist on duty at the time— he testified at the hearings—deemed it unnecessary to advise Flight five fourteen or any other flight of this significant meteorological information. The meteorologist has a very free hand, apparently, in determining whether the information is relevant enough to pass on to the pilots or not—yet he is not a pilot. He has the responsibility for no passengers. He sits there at his little machine and makes the determination."

Like most annual reports, TWA's for 1974 was prepared for stockholders, disgruntled stockholders in this case, and its tone is upbeat, the paper glossy, and the figures carefully selected—omitting, for example, mention of the $11 to $12 million a year the airline spends in hull (airplane) and liability (dead passengers) insurance. Great emphasis is placed on cost cutting. Under a heading labeled "Jet Fuel Conservation Measures" comes the assurance that "we have expanded the use of visual simulators for pilot proficiency training which resulted in a 14-per-cent reduction in aircraft training hours."

Exhibit #2-A submitted into evidence at the National Transportation Safety Board's hearings on the crash of Flight 514 describes how TWA's pilots qualify to land at various airports: "To establish route and airport qualifications, a pilot is required to make an entry at *any* [Italics added] airport served by TWA and classified as regular, provisional or refueling, except at those airports where other means of qualification are approved such as reviewing airport qualification films and *by proximity* of the airport to another airport into which the pilot is qualified.

"TWA has an up-to-date movie of Dulles International Airport." Yet, continues the NTSB's review, "the specifics of each approach are not discussed. The movie states 'Consult your Jeppeson Manual for current information about the specific approach. The movie also states that the highest elevation in the general area of Dulles International Airport was 1480 feet, located 10 miles to the west."

Twelve miles further out, on a straight-in approach to runway 12, the Blue Ridge rises to 1764 feet.

"Qualification into Washington airports may be accomplished," the exhibit adds (paraphrasing the government-approved TWA regulations), "by entry into *either* Dulles International, Washington National or Baltimore-Washington Airport." The highest obstacle near National is the Washington Monument; around Baltimore, a high office building. "The TWA Pilots Route and Airport Qualification Record for the Kansas City East area contains a list of questions concerning airports in this sector. Qualification on the Kansas City East Route is contingent on completing these questions. Question #31 was: 'What is the highest minimum sector altitude on any approach chart at Dulles?' "

Captain Richard Brock, Flight 514's commander, "viewed the Dulles Airport Qualification film on December 11, 1970," five years before the crash, according to NTSB's investigators. They were, however, unable to find his answer to Question #31 on the Pilot's Route and Qualification test. Though he had flown into National Airport twice in September of 1974, and twice the year before, Brock admitted to his crew on crash day that he had not landed in Dulles "in a hundred years." Just what he meant by that, the NTSB did not deem necessary to discover.

First Officer Kresheck had flown into Dulles three months before the crash, but had not landed on runway 12. Whether or not Flight Engineer Safranek had ever touched down there, the NTSB did not mention.

Technically—by the book—the crew of TWA Flight 514 was qualified to land at Dulles Airport's runway 12. Yet the discussion in the cockpit—as the jet nosed down through the clouds toward the ridge—in which Brock refers to the approach plate as "this dumb sheet" chillingly illustrates the captain's misguided self-confidence and reluctance to admit doubt about what exactly was meant by Merle Dameron's approach clearance.

Why, in their indecision, the flight crew never radioed down and asked for clarification (as did the American Airlines captain in a similar situation) was, as Alan Clammer and dozens of pilots put it, "the big question."

At best, one can only guess at the psychological make-up of Captain Brock and his crew, wonder at insecurities, perhaps extrapolate some meaning from the fact that Brock was originally hired by TWA, in 1955, as a flight engineer in the days when FEs were only flying mechanics, not pilots; that he flew as such for twelve years before qualifying for his commercial wings. ("I remember Pat Brock," an aging TWA captain said between martinis late one night at the bar of the Indianapolis Airport Hilton. "We flew in the Connies together. He'd never let the plane go up if it wasn't perfect.") Brock served in the air force, as did Kresheck, but as a jet engine mechanic, not a pilot. His flying time as such for TWA only amounted to 3765 hours; Kresheck, four years younger and his junior in rank, had nearly twice as much. Both based at Los Angeles airport, they were neighbors in Thousand Oaks, California, friends even, thus allaying any suspicion of personality conflicts in the cockpit.

Though there was no doubt about why the plane had crashed—it was flying too low and hit a mountain—there was disagreement among the five National Transportation Safety Board members (two Johnson and three Nixon appointees, only one of whom had any actual flying experience) as to why the crew had flown it down to 1800 feet, or lower, instead of the proper approach altitude of 3400.

In the formal Accident Report, finally issued thirteen

months after the crash, the five Board members concurred on only one point: that after getting Merle Dameron's message that he was "cleared for approach," Brock "did not react correctly to his own doubt about the line of action he had selected because he did not contact the controller for clarification," and that "the crew had at their disposal sufficient information which should have prompted them either to refrain from descending below the minimum sector altitude or, at the very least, to have requested clarification of the clearance. Although the profile on the approach plate did not fully and accurately depict the various minimum altitudes associated with the entire approach, it appears there was adequate information on the plan view of the plate to alert a prudent pilot of the hazards of descending to an altitude of 1800 feet prior to reaching the Round Hill intersection."

While stressing that airlines should "indoctrinate" their crews with the precept "that during flight the final and absolute responsibility for the safe conduct of the flight rests solely with the captain as pilot-in-command, regardless of mitigating influences which may appear to dilute or derogate this authority"—ground controllers' instructions for example —the Board's three-member majority decided that the accident's "probable cause" (the closest the NTSB comes to pointing the finger of blame) was "the crew's decision to descend to 1800 feet before the aircraft had reached the approach segment where that minimum altitude applied," a decision resulting from the "inadequacies and lack of clarity in the air traffic control procedures which led to a misunderstanding on the parts of the pilots and of the controllers regarding each other's responsibilities during operations in terminal areas under instrumental meteorological conditions."

The Board's minority position was that "there was no terminology difficulty. The plain fact of the matter is that the controller simply did not treat the flight as a radar arrival as he should have. . . . The pilot assumed he was a radar arrival and would be given altitude restrictions if necessary. . . . In our opinion the probable cause was the failure of the controller to issue an altitude restriction. . . . The real issue in this accident is not one of inadequacy of terminology or lack of understanding between controllers and pilots. Rather, it is a failure on the part of both the controllers and pilots to utilize the ATC [Air Traffic Control] system properly and to its maximum capability."

Yet, whether the secondary blame lies with the whole air traffic control system or just with the approach clearance given to Flight 514 on the morning of December 1, 1974, by Merle Dameron (who is now training to be a mechanic, having opted for early retirement from the FAA), the report emphasizes that TWA Captain Richard Brock did not react in a way "expected of a pilot suddenly confronted with uncertainty about the altitude at which he should operate his aircraft."

Pilots tend to be a tightly knit, defensive group when discussing a dead buddy's performance. Clammer would only describe TWA's pilots, his colleagues, as "a cross-section of people, like an office. We've got our assholes; we've got some great guys." Shortly after the crash, Dr. Hocker, the Loudon County Medical Examiner received a vitriolic letter from a captain for a major American international airline berating him for telling reporters at the crash site that he would, naturally, perform toxicological examinations on the remains of Flight 514's crew: "To make a point [of talking about such tests] while the bodies are still in the wreckage is grossly unfair to all concerned," the pilot wrote. "It is an ugly suspicion, it plants an unwarranted seed of suspicion in the minds of many who saw your remarks. There are matters within both our professions which are better kept within them."

The tests revealed no trace of drugs or alcohol.

"It's too damned easy to say pilot error, pilot error," fumed another senior airline captain. "That poor guy on TWA five fourteen was caught between a rock and a hard place. The charts and maps and approach plates for Washington are deficient. You come in at National and you need one set. For Dulles just a couple of minutes away, you need another, and for Baltimore still another. You come in from the west and the approach plate only extends out ten miles. Does it show mountains? Hell no. You have to shuffle papers around like crazy as you're coming down. And still fly the airplane. It's the system that's lousy, the regulations, the maps, . . . the FAA and its administrators. Most of them aren't smart enough to empty out a bootful of piss."

Though phrasing its criticisms more elegantly, the NTSB Accident Report listed as contributing factors to the TWA crash, "(1) the failure of the FAA to take timely action to resolve the confusion and misinterpretation of air traffic terminology although the Agency had been aware of the problem [including uncertainty about the specific approach to Dulles's runway 12] for several years; (2) the issuance of the approach clearance [by an FAA employee—the air traffic controller] while the flight was 44 miles from the airport on an unpublished route without clearly defined minimum altitudes; and (3) inadequate depiction of altitude restrictions on the profile view of the approach chart [a document subject to FAA approval] for the VOR/DME approach to runway 12 at Dulles International Airport."

The NTSB also suggested—for that is all it is empowered to do—fourteen specific changes in FAA procedures and regulations, from clarifying terminology, to revamping approach plates, to making it mandatory that all commercial aircraft be equipped with a Ground Proximity Warning Device which, with a disembodied electronic voice, shrieks warnings to the crew if their plane is too low. The FAA accepted six of the recommendations outright, rejected one, and has the others under consideration.

TWA, following company policy—as outlined by their New York spokesman Frank Parisi—not to discuss the accident or crew because "anything we say above and beyond what is in the public record can only hurt us," was mute on the NTSB findings and recommendations. Yet, they too were bucking for changes. Under the heading "Maintenance Programs Expanded," their 1974 annual report said: "We also initiated a review of existing practices and technology in an attempt to maximize efficiency, reliability and savings." Not a word about improving their safety record following a year in which a total of 180 people died in TWA crashes. (Among U.S. carriers, between 1965 and 1974, only Pan American killed more people than TWA: 318 passengers in six fatal crashes versus 250 in five. For the record, Eastern followed with 245, American with 184, and United with 139.)

The closest the report comes to mentioning the two accidents comes three pages from the end of the thirty-nine-page brochure under the heading of "Contingencies, Etc.," which lists, in the following order, the so-called contingencies TWA's attorneys are concerned about: (1) A suit by various landowners whose property abuts airports used by TWA and who are complaining about noise pollution; (2) A class action suit filed by 350 former hostesses alleging they were discriminated against and fired because they became pregnant; and (3) "There exists certain other contingency liabilities resulting from litigation, but in the opinion of TWA's General Counsel, none will result in liability which, over and above any insurance coverage thereof, would materially affect TWA's financial condition or interfere with its operations."

What the report did not tell its stockholders was that an ugly liability trial was in the offing, *Trans World Airlines* vs. *The United States*, to determine whether the crew of Flight 514 or controller Merle Dameron of the Federal Aviation Administration was to blame for the crash, and who, eventually, was to pay the settlements and jury awards stemming from the wrongful deaths of the eighty-five passengers.

The trial was scheduled for December 8, 1975, in the U.S. District Court in Alexandria, Virginia. On November 14, a Friday, the government and TWA jointly announced that they had reached an "agreement." TWA said, "We are not contesting liability." The government said, "We are not admitting it." But, somehow, both were willing to pay a percentage of the damage settlements. The exact breakdown of how much each side was to pay remained secret. "We have restrictions on us by the government," said the lead TWA attorney. "TWA asked us not to reveal the figures," said the Justice Department's Aviation Unit.

"Like any settlement," then confided the TWA lawyer, "it didn't come about until there was a climate of understanding. The depositions which were taken apparently made everybody very conscious of how much exposure there was on both sides. The agreement is purely an economic thing— a practical approach. If the government is willing to share the financial burden of settlement we don't insist that it admit it was at fault."

One insurance man said, "We started working on the government immediately after the accident, annoying the hell out of them. You pound the table quite a bit and, if they're in bad enough shape, they've been known to cough up a dollar or two, but they are not going to take the individual responsibility for X-millions of dollars without pretty good reason. The bureaucracy type of operation lends itself to second-guessing, and you have to go pretty close to the top for concurrence before you pay out a hell of a lot of money. Otherwise, you're liable to get chopped off at the knees. And any experienced bureaucrat knows this all too well. They cannot be in a position of wasting taxpayers' money."

The lawsuits stemming from the crash might be worth between $12 and $13 million, or just over TWA's 1974 annual insurance premium. Were TWA to have gone to trial and been found liable, the entire burden of settling and defending suits would have befallen Associated Aviation Underwriters and Lloyds of London, the airline's insurers. But, in return for help from the U.S. tax dollar (settlements paid out by the government come from the general treasury), TWA was willing to tacitly accept liability.

Compared to the newspaper and television coverage generated by the NTSB hearings, notice of the TWA-U.S. agreement was buried in inside pages and on the mid-morning news. Yet this was the first time—eleven months after the crash—that someone, some organization, stood up and tacitly admitted blame. The NTSB's Accident Report had not yet appeared, but when it was finally made public, in January 1976, it too was given scant attention in the press.

One of the men involved in handling the insurance settlements welcomed the public indifference. "We do the best we can to spare TWA publicity without giving away the store." A crash, he said, is really "a public-relations problem."

JAMES MICHAEL DAVIS, 15, BODY BAG #32

James Michael, as his mother usually calls him, liked the flying part of it as much as anything else. Perhaps more. On holidays and in the summer, he'd ride Piedmont from Norfolk to Washington and TWA from Washington to Indianapolis. From Wanda Sloan in Smithfield, Virginia, to Odie "Bo" Davis in Clarksburg, Indiana—from one parent in one rural backwater to another.

He had not always flown: until he was twelve, his mother would drive him to Charleston, West Virginia—"halfway"— where she would hand him over to his natural father for the long drive to Indiana. Though the drives and the waiting ("Most of the time I'd get there first . . . especially when he was coming home . . . hours that you'd just sit. Nerve-wracking") were far from pleasant, Mrs. Sloan says she preferred them to her son's flying.

But "James Michael wanted to fly," she says. "He felt like they were fussing at me too much when we would meet. Flying was a way of relieving tension."

And tension, Mrs. Sloan explains, is something she has assiduously sought to minimize, both in her life and in her son's, ever since she fled from "the home place" and Odie "Bo" Davis, his temper, and his prize Herefords. "One day you find yourself getting so nervous you're either going to crack up or do something. So you better get out of there. When I got to thinking nasty thoughts like, 'I'll shoot your guts out when you come through the door,' I knew it was time to go."

Tugging four-year-old James Michael along, she visited her sister in South Hampton County, Virginia. The postmaster tipped his hat and the grocery clerk said hello—"everybody said hello"—and she stayed, married again, and, as Hugh Sidey once wrote about another family, settled down to "the enduring business of living, finding fulfillment in family, church and neighbors."

She now lives in Smithfield, population 3000—tidewater country of consolidated school districts and grass strip airports, warm now even in November, with the peanut crop ready for harvest and the stands of pine along Route 460 hazy green in the gentle light.

We meet in the sunshine outside her attorney's office opposite the clapboard railway depot. The plane crashed eleven months ago, but because her suit against TWA is still pending, "Lawyer Baxter," as Mrs. Sloan calls him, is suspicious of people who drive all the way from New York to ask questions. Rumor is, Mrs. Sloan confides, "that you're a front-runner for F. Lee Bailey."

Despite his misgivings Baxter is late, and it is his clerk, a pale blonde girl named Betty, who monitors the conversation. She sits in her boss's leather swivel armchair, supplying us with matches and ash trays and listening, interrupting only occasionally to ask, "Is that reh-levant?"

The legacy of divorce is inescapable. Killed before he could fashion a life of his own, James Michael Davis was (and the memories of him are), inevitably, the creations of his mother. Her recollections are tempered by guilt and by sorrow, by the failure of a marriage which conceived him, all flooding out now in the carpeted and paneled country lawyer's office, a dammed stream of resentment suddenly released.

Mr. Davis will not discuss his son—his attorney, Lloyd Elkins forbids it. Called on his farm in April 1975, Davis first said, "Sure, come on down," gave directions ("Turn left at the silo, right at the 4-H sign"), then called back and said it would have to be arranged through his attorney.

Phone conversation, April 23, 1975:

"Hello? Mr. Elkins? This is Adam Shaw. Mr. Davis suggested that I call you about the book I'm doing—"

Elkins interrupts. "Half the royal-ties. We want half the royal-ties of the book . . ."

"Huh, Mr. Elkins, I don't think you understand—"

". . . half the roy*al*-ties." The word is obviously foreign to him—probably just looked it up. "I want half the roy*al*-ties or I won't talk to you." Voice rising, "Do you understand?"

"Yes, but I . . . I'd like to come down to Clarksburg and talk to you. I think it might be useful to—"

"I said half the royal-ties. The conversation is over. I'm a busy man. We are not going to cooperate. Do you understand?"

The phone is slammed down.

Months and one unanswered letter later (". . . awfully sorry that there appears to have been a slight misunderstanding between your attorney and myself . . ."), Mr. Davis was called again: "Your chance to tell your side of the story . . . have heard Mrs. Sloan . . . You might want to straighten some things out . . .") Again, he is perfectly pleasant but says Elkins must be consulted.

Phone conversation, November 20, 1975:

Preliminary niceties, then Elkins: "My father told me when I was a boy not to get in a pissing contest with the skunk and I followed that rule, and as a result I've kept out of a lot of trouble."

"Huh? I don't think I follow you."

"Mr. Davis is not going to get involved in any damn book. I represent Mr. Davis and he's gonna do what I tell him to do." Pause. Rasping breath. "You have no damned business calling him and bothering him!"

What follows, therefore, are the memories only of James Michael's mother, Wanda Sloan.

James Michael called you from Clarksburg over Thanksgiving?

"This brings back things that can hurt, and I don't know really whether I should say them or not."

Betty is silent. Mrs. Sloan goes on. "Through the years, Mr. Davis tried to tear me down with this child. I don't think he really wanted this child as much as he wanted to hurt me. And that's what they were doing Thanksgiving, because James Michael called me and told me, 'Mama, they will let you know about my schedule later, about getting home.' And I said, 'What do you mean, your schedule later?' He said, 'They only bought my ticket as far as Washington, D.C.' I said, 'Well, how are you supposed to get the rest of the way in?' He said, 'Mama, I don't know.' I said, 'All right, James,' I said, 'I know you can manage; I know you're capable of getting around. Tell them that I cannot come to Washington and get you, and to try and get that other ticket on in for Norfolk.' Why they didn't do it that one time, I don't know . . . I guess it was just making me do things. You know, like, it was against *me*."

Betty looks uncomfortable. "This is really supposed to be about James."

Wanda reacts to this rebuke by talking about her son's activities and interests, his goals and accomplishments. First, James Michael as a reporter for the *Tidewater News:* "He was a little guy that if your team came to play our team, he would find out where you were and he'd call and say congratulations, but he'd also tell you that he was rooting for his own school . . . He talked about being a sportswriter when he grew up. Or a newscaster. I felt like that he had a brain of his own and that he could do anything he wanted to do. I would try to give him the education, and what he liked, what he could do well, was what I wanted him to do. Not what I would have him do."

What would you have had him do?

"Oh, he would have been the President." She laughs.

James Michael as a budding novelist: "He was always writin'," and Mrs. Sloan keeps it all, in boxes on a shelf, in his old room at home—a treatise on success, a short story in which the hero is on dope and drives through Smithfield in a '55 Chevy pursued by the local constabulary—"just dreams."

James Michael as a farmer: "I have a letter James wrote that said his father was going to give him twenty acres of land if he would leave me. I suppose he wanted him to be a farmer by offering him land. He told me about it and I answered him this way: 'If you want to farm, we'll help you get the land. What you want to do, we'll help you with. But you don't have to be mean to anybody or leave anybody to do these things."

James Michael as a ham radio enthusiast: "He had a CB [Citizens' Band] radio that he made a lot of calls on, received a lot of calls on. To anybody, anybody that wanted to talk to him. He wanted a hamburger brought home, he'd get on the CB and somebody would listen and instead of him making a long distance call [calling next door can be long distance because of county lines], they'd call me at the restaurant and say, 'Your son is on the CB; he wants a hamburger.' I was going to Waverly one day and somebody came in the restaurant and told me they knew I went to Waverly. I asked how come, and they said somebody was on the radio —I was Big Mama when they were talking about me—and said Mama said she needed some underwear and I was going to Waverly to buy it." She smiles. "I learned not to tell him everything."

It's time for lunch. We drive to Sloan's Country Kitchen, the roadside café owned by her second husband, halfway between Windsor and Smithfield. On their car is the bumper-sticker "America's Manpower begins with Boy-Power." The café is a neat, bright, little restaurant, Smoked country ham, mashed potatoes, gravy, and corn bread. Mrs. Sloan eats hush-puppies and drinks a glass of milk; Huey, her husband, sits with us, silent but friendly.

Wanda says, "I don't know if there was anybody in the world, including me, that he loved any more than he did Huey—his Father's Day cards, the way he kissed him good-night. The things he told me . . . One thing he told me, just about three weeks before he died: Mr. Davis had showed James a card that he had sent him before he could write. He could print though, and he wrote 'To my real daddy' on that card. I addressed the outside and mailed it to his father. I did a lot of things like that instead of turning him against him. I mailed things, let him know the little fella was thinking of him. His daddy showed him that card, and James Michael was talking to me [later] and said, 'Mama, wasn't I stupid?' he said, 'I sent it to my *real* daddy, and I love Huey a whole lot more than I do him.' He was a kid when he sent the card, but about three weeks before he died is when he told me how stupid he was to write the card that way. And I explained to him that he was just a little boy, that I knew he was just trying to express himself."

Did he have any step-brothers or sisters out in Clarksburg?

"He had a brother and a sister. Billy Randolph and Becky. That's the eldest son. He's a farmer and James loved his little son. That was his big desire, he'd get to see Hank. Hank was the little baby, it's the first baby he ever held. First baby he ever gave bottle to. James was an uncle. He was a little guy and he was an uncle."

It's a fine memory and we smile.

Does Mr. Davis have other children? Did he remarry?

"Well, he's been married several times. There's always children there. I'm not going to go into that. I don't know . . ."

(Earlier, when she was talking about driving her son to West Virginia to be picked up by Mr. Davis, Mrs. Sloan

said, "I got to meet all of his wives." *They would come too?*
"Sometimes." *That must have been pleasant.* "Well, they
were nice-looking ladies . . .")

Irresistibly, she goes on, stops herself, cuts what she wants
to say in half. "The present wife, I think maybe she might be
four or five—I don't know what number she is—she has a
daughter, maybe eighteen months older than James. He liked
her. Her name is Grace, and I put her name as the sister in
the paper when James died. Just because he liked her. But I
explained to the newspaper that it really wasn't no relation."

Such distinctions are important to Mrs. Sloan. They are
her way of establishing the boundaries between her former
and present ways of life—Clarksburg and Smithfield—of
staking out, as much for herself as for anyone else, what is hers
and what is not. Her son; his sons. Her daughter; his brood.
(The Sloans also adopted a son, Henry, whom, she says,
James, aged nine, "picked" out of an orphanage playground.
"Henry thought it was a nice little boy coming to live in the
home. And James Michael thought it was a nice little boy
coming to play with him, but after a while James Michael and
he came in holding hands and James said, 'Well, if we can,
let's take him home. I like him.'") It is a question of al-
legiance, not blood.

Trucks rumble by outside, hauling loads to Norfolk and
the sea.

How did you hear about the accident?

"Mr. Davis called me. And I didn't believe him when he
called, because he was always telling me something. And I
went to pieces and somebody else took the phone over, but I
still didn't believe him. I thought they were just making up a
bunch of stuff to hurt me. Again."

When did you finally know it was true?

"I didn't. I wasn't . . . That happened on Sunday. Mr.
Davis called me about the time I was expecting a call from
James to pick him up. Somebody brought me home [from the
restaurant] and they wouldn't let me hear the radio, they
wouldn't. I put a call through to my daughter to see if she'd

heard anything and she was the one who told me there were no survivors. She was able to announce his birth to me and she announced his death to me. 'Mommy, Mommy, we got a boy and he's got hair—he's the prettiest one in there . . .'

"But it was four days before I was told anything officially. I was angry at first; I was angry at everything. I was angry because it wasn't me. I had lived. Why did it have to be somebody so young, so bright and full of life? And me, I've lived to past fifty. Why wasn't it me?" A small moment for blinking, for swallowing.

"I have accepted the fact that James Michael accepted Christianity. I know where he is and that makes it easier. It also makes me work very hard at being good because I want to play peeky-boo with him again. It makes me a better person.

"It's hard to be a better person when you come to think that perhaps he was murdered. In a certain way."

Through the slanting late afternoon sun, we drive toward the graveyard where her boy is buried.

Davis had called her up, she says, "and threatened to sue for his body. I guess I was torn up and I was pretty mean about that. He called me Wednesday, maybe an hour after I knew that he had been identified, and he told me he was going to see Mr. Elkins and the judge and everybody else. So he could be buried in Indiana. I said, 'Look, I have had him all of his life, he went to school here, he was baptized in the church here, his friends were here, and I'll see you in hell before you'll get him.' "

The day of the funeral Davis and his wife were in the Sloan's house. "My daughter came and got me, telling me her daddy was there. I came out and I walked across the room and I took his hand and asked him would he like to eat—because all kinds of food had been brought in. He had a cup of coffee in his hand. She was sitting beside him and she popped up and just said something about me not letting James Michael drive . . . his wife did. She said something like that and I turned on him and her both, and I told them, I said, 'Look, I have gone out of my way to be good to you. I even went with James Michael and bought Mother's Day cards for you,' I said, 'I wasn't obligated to do that, but I was trying to teach James Michael love. I was trying to teach him that one side could do something right, and I'd go out and pick a Mother's Day card for you, and the very first one you got said "James Michael and his Mama." '

"And Mr. Davis popped up to say something—" four people in a small room, cold coffee on the sideboard, tempers flaring, a funeral about to begin "—I told him, 'You left me when that child was born, I didn't leave you. And you never supported a child, but anytime the child wanted something, well, why don't your mama give it? You gotta run me down all the time.' "

An old man dozes in his mule cart as it moves down the road. On the rise of a pasture, as natural to the land as an oak or a stand of pine, a chimney stands silent guard over the ghost of a mansion. Nat Turner country. It was over these same quiet fields that Nat led his desperate band in a futile attempt at rebellion. And it is in this same, once-rich Tidewater earth that James Michael Davis is buried in the unfenced meadow behind the First Corinthian Baptist Church.

"He used to play football here," she says, "after Sunday School."

MERLE BASIL, 47, BODY BAG #61

Jack Basil gives directions to his house as though transmitting map coordinates for a field exercise—precise, letter-perfect. Basil is in the middle of eating a supper of spaghetti, black olives, and arab bread with his brother. I know nothing about either of them, nothing about the dead woman. We talk about Africa until the dishes are cleared.

"I met my wife when she was a schoolteacher in northern

Vermont. I came to Washington in 'fifty-three and she came
down shortly after that. We continued going out with each
other and in 1955, August 'fifty-five, we were married in St.
Matthew's Cathedral in Washington. We had a good life
and she was a very fine woman—excellent homemaker, a
well thought of school teacher. She was an English teacher—
literature and basic English, grammar, and so forth and so on.
She taught in Prince George's County at Northwestern High
School since 1954. Highly thought of, very professionally
competent. She went out of her way to make her teaching
vivid, to make it vibrant in every way, even in the so-called
dull courses such as teaching certain aspects of English literature, for example, Shakespeare's *Julius Caesar* or *The Merchant of Venice*, and so forth . . ."

Merle Basil had taught in a county of tract housing and
billboards, a neon-infested strip of progress wedged between
Baltimore and the District of Columbia, peopled by those too
poor to afford other suburbs but too proud to live in the
cities—policemen, firemen, GS-5s, and their children.

"She loved it. She had a one-month stint at another school
right by the house here, but she preferred Northwestern, she
preferred the administration better than the administration
at the other high school."

And that, despite having to get up at the crack of dawn.
"Because of busing," says Jack. "Normally she used to leave
here at seven-thirty. Now, in the last two or three years up
until she died, she was leaving at a quarter to seven." Basil
laughs. "Even the Army doesn't get you up that early."

Jack becomes serious. "This busing is a gross mistake, a
grossly artificial means to achieve a basic moral result, and it
doesn't work. I'll bet you that's one question about eighty-five
per cent of the people agree on, that it's wrong. They could
be liberals or conservatives, middle-of-the-roaders, storm troop-
ers—most people think it's wrong. What it does though, it
takes a good quality school and lowers it."

The racial imbalance of Washington must have saddened
and sometimes shocked the Basils, both of whom had spent

their early lives in the bucolic tranquility of Vermont. "She was from the Burlington area and I'm from the northeastern part, Island Pond. I was working with my dad up there. We ran a store. She came from a large family; had six brothers and sisters, she made the seventh one. Her father was an engineer, a civil engineer largely. He built bridges and roads and so forth, a very, very fine engineer too. A very, very nice man. Very nice man. My wife came from a Protestant Anglo-Saxon background, and my background is Lebanese Catholic."

A special assistant at the National Rifle Association's Institute for Legislative Action, Basil helps draft position papers, answer mail, and forecast congressional gun control legislation. He is committed to look at life in a calm, orderly fashion, to consider both sides of a problem and come up with reasonable answers, in his work and in his life.

"While we [at the NRA] may be somewhat one-sided, as all organizations are because they have a particular position to espouse, we do have our own approach to arms and controls, and we're not wedded to any particular direction or orientation . . . The NRA's been getting a lot of heat for quite some time. It doesn't matter what comes up, I mean the heat doesn't change in substance, it just changes in degree. And very frequently there's a lot of heat but not too much action. As you know, with lawmakers, for various psychological, social, and other reasons, they'll grab onto an issue and talk about it till they beat it into the ground. But they really don't have a deep-down desire to come up with any specific, concrete action. But I'm not saying that's always the case."

About the plane crash he says, "Well, I think it's one of those things that happen. How do you . . . I mean, it doesn't matter what intellectual or moral process you go through, the fact remains that it did occur, that so many lives were lost and that's it. You try to do the best you can. It's too bad though that so many fine people, people like my wife have to go so early. My wife was only, what, forty-seven years old when she died, making a highly, a very significant contribution. Very significant, both in her private life and her professional life.

"I think of her very often . . . I might engage in a few more activities than I would have engaged in as frequently if I were married because of other claims and demands and so forth, but I'm very busy. I myself, I'm an army reserve officer, commanding officer of this headquarters so I have fifty officers and about one-hundred-and-some enlisted personnel under me. I've been in the army reserves since 1949. I was in World War Two for approximately three years. I was an enlisted man and was stationed in North Africa and Italy and spent about twenty months overseas and it was good experience. I also went on active duty during the Berlin crisis; our unit was called up at that time and we went to Fort Gordon, Georgia. I was a major then on active duty. I then spent three months in Berlin as planning officer, and my wife joined me after my tour in Berlin. I had her fly over and we spent thirty days touring Europe. We rented a Volkswagen and went through six or seven nations. It was her first time and she enjoyed it thoroughly."

Theirs must have been a meticulously organized, not at all unpleasant life; hard work rewarded by good things.

"She was a very fine dresser, dressed very, very nicely, clothes and shoes and so on, the closets attest to that, yeah . . . She received innumerable catalogues and magazines from the finer stores and shops in the United States, she was particularly interested in the specialty shops you know, that cater to this particular thing . . . When I got my promotion to full Bird last year, to full colonel . . ."

How does a promotion work?

"There are several factors involved. Position, educational background, experience, time, so forth. The modern military officer today is a highly competent, a highly educated man, both civilian-wise and military-wise. In fact, they expect the army and navy officer to go to school for a good part of his life [Basil has a Master's degree in Political Science from Georgetown University, gained at night], he goes to various staff schools, which are excellent, like Command and General Staff and various war colleges."

You were promoted to full colonel . . .

"Yes. And Merle was supposed to pin the eagles on me.
She had bought the set of eagles and was wondering week by
week when it was coming in. I got the promotion—paper-wise
it was effective from June 'seventy-four, but the actual promotion itself didn't come in till the latter part of November.
The ceremony—because of the funeral—was postponed until
January." Not an insensitive man, Jack Basil pauses, saddened
both for his wife and himself.

Thanksgiving, shared in Columbus with Merle's brother
and his family, was "one of the finest," Basil says, "extremely
pleasant," marred only by his early departure, on Saturday, to
attend a convention in Houston. Merle boarded Flight 514 to
go home on Sunday.

"I was at our booth in Houston," Jack says, "when somewhere around noontime I heard about this air crash . . . So I
came back that night, and I returned to Dulles." The Basils
had no children. Jack walked into the house in Silver Spring
alone. "In my little den, I found her sewing kit with her dress
draped over the chair ready to be worked on."

Now he hunts for mementos: awards, high school yearbooks, In Memoriams . . . "That's it. The things are reduced
to newspaper clippings and little articles." Again, he will not
let himself give in. His voice brightens, becomes louder. "Very
interestingly, I've got . . . you know, they retrieved from the
air crash a number of jewelry items and so forth. Even though
part of the bodies were mutilated beyond recognition, certain
items are absolutely intact. One of them is as if she had just
handed it to me. Absolutely amazing. As if she just . . . I've
got her wedding ring, her checkbook—this I had to look over
—she even had her last balance in it. And this is her credit
card holder . . ." He handles the cards, the little brown leather
wallet gently, but no longer with awe. The tangible legacy of
Weather Mountain. "Look, she even had them in alphabetical order, see, see? In alphabetical order, the last things. Everything was meticulous. And I have many other things, but I'm
going to put this back in the safety deposit box now."

He comes back. "She had little notes . . . books she wanted to give to her class, a little sketch of a cabinet . . . And her scarf, a little head scarf. You can still smell the perfume on it."

JOHN A. DEPEW, 70, BODY BAG #9

"We didn't have a funeral because, you see, I got nothing. We had a memorial service for John . . . He was killed on Sunday and we had it the following Saturday. The next day we had a dedication service at the church. We were dedicating the new building and the pews and the Bibles. Normally, I wouldn't have gone out in public then, but I did because John and I cared so much about this. So, when this girl brought me home, and she didn't want to leave me, she said, 'Mildred, this is the first time you've been alone.' And I said, 'Great. I want to be alone. I've had company, somebody in this house, for a week.' So she left.

"It was about quarter to five in the afternoon and it was snowing like mad. One week from the day it happened. There was a knock on the door. There stood a young man who said he was a lawyer from TWA. He handed me a card and there was a name—not his name—on it and then, in pencil, was written TWA . . . in pencil. Anyway, he made this statement: 'Your husband was seventy wasn't he?' 'Yes,' I said, 'he was.' And he said, 'Well, you know . . .' In other words, your husband had nothing to give any more, he was over the hill. I said, 'Look, my husband had more on the ball than a lot of guys that are fifty. Don't you dare say anything because my husband was seventy years old!'

"I knew he wanted to make a settlement with me."

Mildred Donovan and John DePew met after choir practice the spring America went to war in 1917. John, whose father served as postmaster for the town of Zion, Illinois, was eighteen years old and worked as an upholsterer. Mildred, the new belle in town, a minister's daughter, was fifteen and wore her hair bobbed. Bashfully, tentatively, they courted, separated, and, a decade later, were married. America was dry and, in other places, gay. Scott Fitzgerald called it the Jazz Age and wrote that "in that spring of '27 something bright and alien flashed across the sky. A young Minnesotan who seemed to have nothing to do with his generation did a heroic thing, and for a moment people set down their glasses in country clubs and speakeasies and thought of their old best dreams." Lindbergh had landed. Everyone was humming "Bye, Bye Blackbird." Nicola Sacco and Bartolomeo Vanzetti died in the electric chair.

During the Depression, John DePew sold office supplies, begat two sons, and, at the age of thirty-six, decided he wanted to become a state policeman. He was turned down at first, but he persevered and became a trooper. "He patrolled the highways," says Mildred. "He rode a motorcycle. They had the puttees then, the tight pants, and the Sam Brown belt . . . I passed him one day in a car—or he passed me—and later I said, 'Honey, were you as nervous as you looked?' 'Every bit,' he said. And John never knew what a bribe was. One day a girl said something to me about 'You and your state police husband—all those bribes and everything,' and I said, 'Now look, Flossie, you have a son in the navy, haven't you?' She said, 'Yes.' I said 'All right, what about these girls in every port?' She said, 'Well, that doesn't apply to Ralph. How dare you say such a thing?' I said, 'It doesn't apply to John either.' "

After World War Two, John DePew left the state police to serve as safety and personnel director for a trucking company in Rock Island, Illinois. Refusing to remain idle when the company retired him at sixty-five, he moved to Indianapolis and took a job as safety consultant with a truck driver training outfit, in his spare time testifying as an expert witness in court cases and lecturing on traffic safety. "We moved here four years ago last January," Mildred says, "but I cried for a whole year . . . This man would come home from work every night and all I wanted to do was go home. I wanted to go back to Rock Island. I don't know how he ever put up with me. It was sad. I gave him an awful bad time."

The DePews lived in a tidy second-story apartment in Speedway, not far from the Pierces (whom Mildred met after the crash), and were firmly ensconced in the pleasant routine of an old and comfortable marriage—a drink or two every evening (Martinis for John, Bloody Marys for Mildred); a meal out at a nearby restaurant; occasional trips together ("I went with him; I either had to do that or I wasn't seeing him"); nightly phone calls when Mildred stayed behind; and, lately, the satisfaction of two grown sons leading decent, productive lives of their own.

DePew was going to Washington on December 1 to attend a meeting of the American Transport Association Council of Safety Supervisors, of which he chaired a committee. "When he got up that morning it was bad, so I said, 'John, why don't you call and see if they're flying.' So he came back and said, 'Yes, they're flying.' And he said, 'I called last night and ordered my cab for a quarter after seven. By the way, what are you doing awake?' I said, 'Oh, I don't know . . . I think I'll get up and get you some coffee and juice.' So he went to our storage here to get me my boots so I could go to church. When he came back in he said, 'The cab's out there.' It was twenty minutes *of* seven. They thought he meant quarter of seven instead of quarter after. And he said to the man, 'Don't toot your horn. I'll be out.' So he said 'I won't have time to have the juice and coffee, and here are your boots.' So he put on his overcoat—he had a black cashmere overcoat, that's John's mark, his trademark, that's how they knew John DePew, his clothes—he put on his overcoat and I said, 'Honey, there's a button missing off your coat.' I asked him when he had last worn it. 'The last time was when you and I went out for Thanksgiving dinner,' he said. He'd already taken his suitcase down so he said, 'I'll promise you something—when I get to Washington I'll have that button sewn on.' And I said, 'Oh please do because I just can't have you going around Washington with a button off your coat.' And he also wondered where his galoshes were. John had a pair of gray and black lizard boots and I said, 'John, why don't you wear your boots? Then you won't have to take your galoshes.' And he said, 'That's a good idea.' So he put his boots on and I saw that everything was all right, and that man walked out . . . I went to the top of the stairs—we had snow about that deep from the steps—and he walked out that front door and out of my life."

Eight months after the crash, Mildred moved out of the Vinewood Drive apartment and into Westminster Village, a retirement home on the northeastern outskirts of Indianapolis. A lucid, correct woman in her late sixties, she now lives in two small rooms surrounded by widows and a handful of widowers ("We have ten bachelors here . . . but, no, that part of my life is, of course, completely over"). She moved out, she says, because she could no longer bear the loneliness and the memories of the apartment she had shared with her husband for nearly half a century. "I would wake up in the morning and there was that empty bed and that would sort of start the day off bad. In the afternoons I'd just die—waiting for that door to unlock. I got so I couldn't eat at the dining room table—I couldn't eat. I ate in the kitchen. I couldn't sit at that table and see that empty chair. So I tried sitting in John's chair and looking at my chair. It still didn't work. And when I walked out of that apartment one of my neighbors said, 'Mildred, we watched you—you didn't turn around and look back.' I said, 'You bet I didn't.' I cut that cord. That life was over. It was great but . . . I knew I had to have a different kind of life, a new kind of life. I shed many, many tears but I'm not going to do it anymore because it didn't change a thing. I have to accept it. There have been a lot of people who have had to accept a lot worse."

And money, mercifully, is not a problem. Mildred was able to collect $25,000 from her husband's American Express card automatic life insurance policy (as did the families of six other passengers on the plane who knew enough to call the company and demand the money. As one Amex spokesman said, "If they bought their ticket by using their card, they were entitled to it. But they have to call to collect") and, after much delay and haggling, a settlement from TWA's insurance company.

Mrs. DePew had an attorney who said he was going to ask for a quarter of a million dollars. She accepted that figure as reasonable but, she says, "One day a man called me up and said he worked for my attorney, Paul Brenner, and he said, 'Mrs. DePew, I just feel that I have to straighten things out.' I said, 'Such as?' He said, 'You're not going to get anywhere near what Mr. Brenner told you.' I said, 'Well, who are you to call me up and tell me this? What happened to Mr. Brenner?' I hung up."

A few days later Brenner and the man who had phoned to tell her of the realistic appraisal "came out and talked to me for two hours. And Brenner didn't open his mouth. He sat there like a little kid and when I said something about the two-hundred-and-fifty thousand dollars he said, 'Mildred, are you sure that I was the one who told you that?' I said, 'Mr. Brenner, you know that I didn't talk to anyone else in that company but you.' So then this other man says, 'Well, I'll tell you what we're going to do, seeing that John was seventy' and all this and that, 'they have offered you sixty thousand dollars.' I said, 'I won't take it.' He held up a paper with figures on it. I said, 'That's nothing but an arithmetic problem for you. Will you please turn it over and put some humanity into this. My husband did not die—he was killed! They took his life and ruined mine.'

"Anyway, I knew I was in for a battle. Then the attorney called and said they'd give me seventy-five thousand dollars. I said I wouldn't take it. Then he said he had called a lawyer in Virginia who said he could get me one-hundred-and-seventy-five thousand dollars if I wanted to wait two years, and that I might have to go to Virginia, to go to court. I talked it over with the boys and they said, 'Mom, don't do it. It isn't worth it. We want you to get out of it, to get it over once and for all.' So then the lawyer finally called again and he said, 'Mrs. DePew, I got eighty-thousand, but believe me, take it because I can't get you anymore. I tried awfully hard to get ninety so we could give you sixty and we would have taken thirty,' but he said, 'If you don't take this, they'll put your name at the bottom of the list and they'll pay you when they get to you, and maybe there won't be anything left.' So they sent me fifty-seven-thousand five-hundred dollars and they kept twenty-two-thousand five-hundred. So I thought it was pretty darn good."

ORVILLE K. ROE, 55, BODY BAG #75
JEAN ROE, 55, BODY BAG #60

Mr. and Mrs. Orville K. Roe of Lafayette, both 55, were en-route to a convention in the Washington [D.C.] area. Roe was an engineering assistant in the Tippecanoe laboratories of Eli Lilly & Co. He had been with the firm 25 years. He and his wife, Jean, had lived in Indianapolis 20 years before moving to Lafayette seven years ago. Survivors include their six children.
—OBITUARY

Built on and around a series of knobby hills bubbling up from the surrounding plain, Lafayette, Indiana, looks like it belongs in West Virginia: neon signs burning in broad daylight, grand marquees for decaying movie theaters, Army-Navy surplus stores, men wearing Bermuda shorts to work. It is August.

"Lilly's? You can't miss it . . ."

The cop is right. First there is the sign, remarkably modest and in need of a fresh coat of paint, and then there are the buildings themselves, brick and concrete, connected by tubes, pipes, and walkways—a clean oil refinery—all surrounded by a closely woven chain link fence topped with barbed wire.

Dr. Albert W. Hubert offers a slightly moist hand to shake. He is Orville Roe's former boss. We walk up a flight of echoing stairs to his office in the administration building. The office is not large, overlooks the parking lot, and has a blackboard on one wall.

"You want to know about Orville? . . . He had a degree in chemical engineering from the University of Nebraska. We hired him in 1950. First he came to work for us in the Indianapolis Group, evaluating chemical and engineering schemes as pertained to agricultural and pharmaceutical production. We designed equipment schemes to commercially produce new products developed by the lab. He was in process research and development. Then Orville came up to Lafayette to work in the production process, to make sure it was being run in an optimal way.

"Let me put it this way . . . Orville was not an applied scientist making the initial discoveries. He was the fellow who figured out how to economically and consistently *produce* the products the scientists discovered. He was the guy who knew how to *make* them."

As a chemical trouble-shooter Roe, Hubert says, was dispatched to Lilly plants around the world to juggle formulas when products' production costs rose and their yields dropped. "For example, he was sent to England to get a hypoglycemic compound back on track—no, I don't want to get too specific on that . . . He was an expert on twenty to thirty processes. How he did some of these things we'll never know. Some were on paper, some in his head. Say there's a chemical we only have to make in quantity once every three years, Orville was the guy who knew how to make it, easy."

In another age he might have been a craftsman, a man with a gift for tinkering, who might have been piecing together caned chairs or muskets. "He was a little like an artist, he had a feel for these things." Yet, as a man "he was very conservative, maybe a little tight with his money . . . though of course he was more than adequately compensated for his work, more so than a full college professor." (The inference is that this was so though he did *not* have a Ph.D., a social and professional commodity which, in the world of Lilly and Lafayette, is akin to belonging to the right country club.) "He had a real soft heart underneath it all which he was always trying to hide by a kidding veneer. He was . . . a little loud sometimes. He liked horseplay, but he was a real soft bundle of mush underneath."

Hubert tells what must be the favorite Orville Roe story. It seems that Roe once tiptoed up behind a lab researcher engrossed in his work and clouted a tin garbage can lid with a broom handle. "The guy from technical services," Hubert says, "almost crapped in his pants." Hubert smiles indulgently at the memory of the incident. "I said to Orville, 'Aren't you a little old to be doing this?'" He says he does not remember Roe's answer and adds, "The company makes a man's box big enough so he can have an input on creativity. Our philosophy is that our greatest assets are our people," who, he is happy to point out, are not nor are allowed to be, unionized.

"Orville," he says, "was always treated fairly here. He seemed pretty well fulfilled; he didn't feel he'd been screwed or anything. Let me show you." He goes to the blackboard and draws a big white cross. In a loping scrawl he chalks the words *fulfillment* and *despair* at opposite ends of the vertical line and *failure* and *success* at the ends of the horizontal arm.

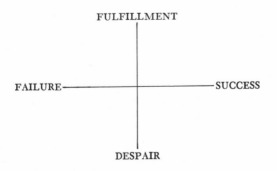

"This," he says, "is how I like to represent life: Lilly-life versus total-life." Obviously this is the diagram he uses when junior executives come in to talk about "my future with the company."

Still standing by the blackboard, Hubert says, "Let me tell you a little story. I knew of a girl who had all the good stuff." He points to *success*, next to which he writes *title, power, $, Cadillac*. "And one morning she woke up, or rather didn't wake up. There was a bottle of pills, empty, by her bed. Her name was Marilyn Monroe." Pointing to *fulfillment* he says, "Orville always worked at this level. He was fun and had fun because of his value system." The pointer draws an imaginary line in the upper right-hand quadrant of the diagram, between *fulfillment* and *success*. "He achieved a balance."

Hubert sits down. "Orville," he sums up, "was perhaps a little square—" he makes a square with his fingers—"but then, Lafayette is in a county that doesn't really know how the world really lives." Except of course for "the pseudo-jet set, the people with private planes who go to Chicago to shop or for dinner."

It's the end of the working day. Hubert and I leave together and shake hands in the still heat of late afternoon. He drives a powder-blue Cadillac and waves goodbye.

Tonight there is to be supper with three of the Roes' sons —Jerry, Theron, Orville Jr.—and their sister, Jane, at the family house down Lilly Road. The other two sons, Jon and Peter, are in Rhode Island. "It'll be an excuse for a little gathering," Jerry had said the day before, "and we'd like you to see the place. Mom and dad built it. Seven o'clock okay?"

I have time to waste. The nearest bar is in a bowling alley. It doesn't serve hard liquor and is crowded with farmers, grease monkeys, and truck drivers who sip beers and watch the CBS evening news on a color set above the bar. Connie Chung is reporting a story on Nelson Rockefeller. The men at the bar whistle. One of them says what all of them are think-

ing, "Boy, would I like to get my hands inside her britches!"
Seven o'clock takes a long time coming.

Orville, Jr. ("Orve"), answers the door and we go into the
vast living room. The blinds are drawn against the setting sun.
He says he's just come back from Europe, so we talk about
that for a while. He had a Eurailpass; he liked the Louvre
in Paris—especially the *Mona Lisa*—and the Impressionists
in the Jeu du Paume. He is wearing red sneakers and a football
jersey. He's twenty-one.

Jerry, his wife Dixie, and their son Matthew arrive. Jerry
lugs in a cooler full of Coors. He's wearing shorts and is full
of cheer. Theron, chronologically between Jerry and Orve,
comes quietly downstairs. He's dark and slim, wears faded
cutoffs and a tie-dye yellow T-shirt.

The three brothers insist we take "the walk" around the
property. "Mom and Dad did it almost every night."

The boys are proud of the wooded, overgrown and un-
spoiled land. A path leads down and away from the back of
the low house toward the Wabash River which marks the
northern end of the estate. Across the river, summer-low and
muddy, and beyond the first set of fields, is Purdue University,
which both Theron and Orve attend. The walk is perhaps a
mile and a half. We get back at nightfall sweaty and covered
with burrs.

Jane—the boys' younger sister—and Dixie have made
spaghetti from Mrs. Roe's recipe. We eat in the kitchen and
drink iced tea. The spaghetti is good. Janie takes a little rib-
bing from her brothers, mostly Jerry, about a boyfriend. She's
blonde, athletic, very all-American. Theron seems embarrassed
by his older brother's open heartiness. Orve shyly eats his food
and remains silent. It's unclear whose lead he'll follow—
Jerry's middle-American-boy-scout-leader, early-marriage, beer-
drinking track, or Theron's more sophisticated "I'm working
on getting my head straight" approach.

After the meal, the brothers, Dixie, and Jane exchange

presents. It's been a lot of people's birthdays, including Matthew's. Theron leans on a doorframe and leafs through bills
—the running expenses of a house that is suddenly theirs.

Dixie will stay in Lafayette, but Jerry needs a ride back to
Indianapolis. We drive south through the warm night. He
talks of his father: the youngest of eleven children born to an
Iowa farmer bankrupted by the Depression; college in Nebraska; married at the onset of the war; Army Corps of
Engineers in the Pacific theater; then Indiana and Lilly's
for a quarter of a century. "Lilly's had an anniversary party
for Dad in August and all of us kids except Jon went. They
kind of roasted Dad and we got a different insight of him, of
how his peers looked at him. Of course it was very flattering
to him, but it was something I'd suspected but which I'm sure
the younger kids had never seen . . .

"He loved his work and I think it affected all of us, his
working there, because he loved to go in in the morning, and
that's an attitude that stays with you. He had a couple of opportunities for advancement or to change over to research
but he didn't do it. I will always consider it beneficial having
known him being happy with his work. It affects your whole
values. He was a strict believer in 'Do what I do, not as I say.'
He would never take a bar of soap from a motel, not because
he wouldn't, but because we might see him. Though most
adults are aware of a gray area, he tried to teach by example
with black and white. He smoked, but I never saw him. He
and mother kept a carton of cigarettes on the top shelf above
the refrigerator, but they never smoked in front of *me* . . .
age twenty-seven."

Jerry admits to being the black sheep in the family. "I
had my good years and my bad years. I had times when they
were very proud of me and I had times that they were more
disappointed in me than they could have ever imagined . . .
Academia really bummed me out. I couldn't hit [college] four
years in a row. I was in elementary ed., which was okay. I was
doing shit like kiddy-lit . . . had to give two hundred and fifty
book reports on children's books. I enjoyed the books but

didn't like writing the reports, things like the social-emotional implications in 'Peter and the Wolf' ... Even when I was in high school, at Shortridge, it was uncomfortable to be around Dad alone because I knew my shortcomings and I'd rather not talk about them."

He suggests that we stop to buy some beer. A six-pack of Stroh's. Back on Route 52. Jerry plays the radio. Nightride. Free associating ... "My biggest grievance with Dad as a youngster was ... At home there wasn't a thing you could do to make him mad, but in public ... Like if you went on a picnic with another family, and this was my perspective as a seven-year-old, then he was always trying to get you to pay. You know, don't swing too high, don't make so much noise ... He wasn't that way except when he thought his kids were imposing on other families. This is silly but like if you'd go out with another couple and they'd each take two lawn chairs for the husband and wife, if one of his kids would sit in them, he'd evict them immediately, he'd say, don't get in Mr. so-and-so's chair. It embarrassed me a lot. He was overprotective about how his kids behaved in public ...

"We used to park cars in our driveway when we lived in Indianapolis—we lived right across from Butler College's football field—people that were going to sporting events, one buck apiece. One time I came home—I was working at scout camp in the summer—and we were having a late dinner and Peter and Jane were probably four and eight years old and it started raining. Peter came in and said, 'Dad, it's raining,' and they ran downstairs and got a bunch of old newspapers and took them over to the Starlights Musical [on the football field] and sold them to people to put over their heads. Dad was an entrepreneur in a small way ... We had a small acreage down near Spencer that we used to go on weekends, we'd camp out down there as a family, and plant Christmas trees. Dad would always take Easter to plant trees and then at Thanksgiving time we'd cut them as a family and sell them at Christmas to our friends ... He was an entrepreneur and it didn't bother him a bit that his kids were over there hustling with the papers. I was a little embarrassed. I was a teenager and I said, 'Jesus, Dad, somebody's gonna see them.' And he said, 'I don't give a shit.' Except he never cussed; he didn't even say damn."

Jerry reflects on the small contradictions of his father's life, opens another beer.

Softly, "They never did anything without us. Mother had some friends that used to always leave their kids with somebody else and go on vacation, but they would never think of it. Dad always said he wanted to go back to Australia [where he had served in the wartime Navy] with Mother. But the big thing he did that summer, he took the kids West. So if he'd had to choose—he'd do everything he felt like he wanted to for someone else. The Australia thing would probably never happen because he just wanted that for him.

"Mom and Dad were very sacrificing people. We didn't know that till after they passed away, and they didn't need to be. Dad was prudent but not to the point that it hurt anybody. Like we drove a '49 Chevy till '57, then he bought a '55 Pontiac which he used till '63, but the whole time he was putting . . . See Lilly's has a payroll bond issue thing where they match fifty cents on the dollar and you're allowed to put up to six per cent of your income in it, and at the time we thought we were poor, Dad was pouring six per cent in because it was being matched . . .

"It's funny, we all thought we paid for our education. Like I had a savings account and a checking account, and I had about four thousand dollars in the bank when I started college, and I'd write mother or call her and say, 'Hey Mom, I'm running low; would you go to the bank and transfer some money?' and sometimes it would come out of my savings and sometimes it wouldn't. She'd just throw in a couple of hundred bucks or whatever. The point was that we all thought we were self-sufficient, but in actuality they helped. They wanted to make us self-reliant because we, none of us, would have *asked* them for money because we didn't think it was fair. We just wouldn't have. When my wife and I bought our house, after great deliberation, we asked Mom and Dad if we could borrow some money for the down payment, a very nominal sum—three thousand dollars, which is pretty nominal—and we knew they'd sold a farm in Nebraska and . . . we deliberated for a long time and finally asked them. They said, 'Well, sure,' and we wrote up a legal loan. Well it wasn't . . . probably a lawyer would have had fits, but we wrote up a note to each other, interest and the whole bit, and then Dad said, 'Now, whatever happens, don't let this worry you because I don't want to lose a son over a loan.' So it was a very workable relationship, particularly when I became unemployed."

Long swig of beer.

"It's a very sad story . . . I gave four and a half years to Western Electric. I was an installation man. I worked nights most of the time, some seventy-hour weeks, and then I was laid off chronologically . . . because of the economy. When I got laid off, I was working as a working engineer, or supervisor. It wasn't a white-collar position but I really enjoyed my work, and the guys that I worked with, my bosses, they all said they had like thirteen-year men that were minorities that couldn't tie their shoes but whom they had to keep. Then the telephone company wanted me to come and work for them and I couldn't because of the affirmative action program. So they put me in a woman's job. It was all clerical and it was a bitch. I'd been there two weeks and ATandT called and offered me a job at eighty bucks more a week and I said, 'You bet!' Then they called back the next day and said, 'The phone company claims they have a training investment in you,' and that they wouldn't be allowed to take me. I'd been there two weeks . . . and it was terrible because they finally ended up asking me to resign. They'd needed someone to fill up with their quotas until November, then November rolled around. I refused to resign and they terminated me.

"And during that time my little boy was paralyzed. We were at a baseball game the night after I got laid off—the Indians were playing Omaha—and he got hit in the head with a foul ball. He was unconscious for about fourteen hours and in intensive care for a couple of weeks. He was almost a year old and he couldn't move his whole left side. The only thing we could do was look at him in the crib."

Matthew is fine now but 1974 had not been a good year. "I had that awful job and then no job and then Matthew got hit and then Mother and Dad passed away and then my wife miscarried..." In the headlights, signs to Stockwell, Clarks Hill... "It all kind of hit at once."

He, Dixie, and Matthew were in Ann Arbor visiting old friends for Thanksgiving when they learned of the crash.

"We'd celebrated Thanksgiving at home a week earlier, my family and my folks. It was my birthday so we had a very nice visit plus, in retrospect... It's a small thing but Mother had given me a check for my birthday, which she generally does, a very nominal amount, but I took the time to go in and thank Dad. Usually he doesn't know until it comes through the bank. It's not that he doesn't care, but Mother does it. I went into the kitchen and I said, 'Dad, we really appreciate this.' At the time I was unemployed and I said, 'I wanted to thank you, I knew you'd know I'd appreciate it, but I wanted to tell you that we really enjoy this, because during unemployment it's nice to know that you're there.' So I was glad that I'd done that, because I'd never done it before...

"Anyway, we were [at our friends' in Ann Arbor] watching the football game, and it came on TV. They said there'd been a wreck and there was even a little levity about it, you know. I turned to my friend and I said, 'How would you like to be the guy that sold people tickets for that plane?' It didn't even occur to me that Mom and Dad were on it. We'd even forgotten they were going to Washington." For a convention of the American Institute of Chemical Engineers. "Mother was gonna do Christmas shopping and they were gonna visit with Dad's sister in Maryland. It was the first trip they'd had alone together in a long time. We didn't even think about it and it was like three hours before I got the phone call . . . Orve was trying to locate me and he'd called my friend's father in Indianapolis who, in turn, called up to Ann Arbor and said, 'I think you ought to call your brother . . .' "

So Jerry rang the quiet house in Lafayette. "When I talked to Orve he hung up like three times, he was so choked up. He couldn't talk, but I needed to get some facts, you know, 'Who's there? Are you all right? Does Grandma know?' He'd have to hang up and then I'd talk to him a little bit later . . . and the last time I talked to him in the course of about an hour, the minister was there. Orve had to get medication—he took it very personally because he put them on the plane."

Colfax. Jerry's on his fourth beer. The empties clink on the floor.

"The Interstates were closed so we spent the night in Ann Arbor. We got out of there about noon and got to Lafayette the equivalent of twenty-four hours later. It didn't hit me till I got home, until I got to the house, till I walked into Mother's and Dad's house. Until then I felt driven just to get there—Orve, you know, the immediate needs, not my personal loss. But as soon as I walked in the house I had to go to another room . . .

"My whole feelings were sorrow for Mom and Dad. 'God,' you know, 'they're never gonna see Matthew grow up, they're not gonna see Jane get married'—Mother's only daughter, which she wanted for forty years—and that was my whole thing, what they were gonna miss. Then, days later my thoughts of that came around . . . because Mom and Dad were so happy and I'm confident that we'll see it. So . . . my grief didn't turn personal for a long time, you know, maybe a month. Mother's Day, which I guess is in May, was the first time I really felt bad again, because my little sister spent it with us and I hated, I just couldn't let her go home to an empty house."

The house?

"Well, there's a tremendous amount of upkeep in this

type of operation . . ." Jerry feels they ought to sell it. Theron, Orve, and Jane still live in it. Orve would like to keep it. The settlement money will be coming in and will help.

Who makes those kinds of decisions?

"We run the family by committee, which is probably not the most effective way, but our . . . the biggest problem with our committee is that no one wants to say anything; we're overly cautious not to step on anyone's toes. So it's, 'Well, I don't care, what do you want to do?' As opposed to one of us being more . . ." Jerry lets it trail off.

Since his bout with the telephone company and unemployment, he's been working for a new office complex, in charge of readying the building itself for occupancy: a drill sergeant bullying contractors, maintenance men, elevator companies, air-conditioning technicians. He's used to making decisions, and rather enjoys the process.

"Probably the biggest problem has been the decision the youngest two, Peter and Jane, have had to make. We said to Peter, who was thirteen, 'Well, do you want to live with us, do you want to stay as you are, or do you want to live with Jon?' who is a teacher at a private secondary school in Rhode Island, a boarding school. It's very famous, it's called Moses-Brown. Anyway, that's a pretty big decision for a thirteen-year-old to make in a week." He decided to go live with Jon.

"Jane knows pretty much what she wants to do, or knows pretty much what she should do. See, there's . . . I'm gonna raise Matthew different than Mother and Dad would have raised him, and with Jane or Pete, you have to make . . . There's a point of disembarkation, you know. Are you gonna . . . is there any reason to raise her like Mother and Dad would have, which is something we couldn't do, and wouldn't try to do, or . . . But it's something that we're aware of. Like, who can she go out with, how late can she stay out at night."

The brother as a parent.

"Well, we look at her as a mature woman, and mostly we look at what Mother and Dad taught her. The principles are still there, but you could get really hung up if you tried to simulate what their guidance would have been with someone else."

We pass Hazelrigg. Sleeping Herefords, a road sign. Soft prairie night.

"After [we reached] the age of eighteen, Mom and Dad never really asserted anything upon any of us. In fact, in some cases, I know Dad went out of his way not to, and I was upset later because I wanted his opinion and I couldn't get it. In retrospect I'd say, 'Well, Dad, what would you have done?' and he'd say, 'I'd have done it this way,' and I'd feel, 'Why the hell didn't you tell me while I was doing it?' and he'd say, 'Would you have listened?' My answer was 'Probably not.' And he'd say, 'Well, that's why.' He was a very mellow person in that respect. It's not very many people that can watch their children make mistakes. It's kind of like letting them pay the tuition for the learning experience. Dad was good about that.

"He seldom took an adversary position, even to things he didn't like. He'd lose his temper if you spilt your milk, you know, a flash of temper, but to actually be contemptuous toward anything or anybody, no. Only if he saw something as a continuing cause for frustration to somebody else, you know, like a political guy—you know, why doesn't that guy quit prosecuting . . ."

Like Watergate?

"It was something they felt very bad about . . . betrayed sort of. I suppose they would have voted for Nixon. I'm pretty sure they did. They knew how they voted and they did it for a reason—it wasn't that they just pulled a lever—but they didn't think it was anybody else's business."

Yet both his parents were joiners, he says. "Mother was the head of about five things, like an organ committee to buy a new organ . . . She'd never done this when we lived in Indianapolis, but here she got stuck into the YWCA somehow, for inner-city instruction for kids. It only took them about

three months to get on every sucker list in town. Dad was a
scoutmaster when he died and it was kind of funny, 'cause I
was too, and Dad was still sleeping on the ground [on scout
outings] and I'm starting to think I'm too old for this shit
and there he was, still doing it. He'd come in Sunday night
from a weekend and just look beat. But he'd do it without a
question, without even questioning himself . . . I don't think
there was anybody on that plane that left a bigger hole in the
community.

"At the services for them there were about seven hundred
people there . . . The problem was confirmation of death, so
we went ahead and held the services without remains. They
said we haven't found your folks' bodies yet . . ."

Does it matter . . . the bodies?

"Well, yes, to a minor extent. One of the first things is
denial. I mean, you do that automatically and you say, 'Well,
maybe they got off in Columbus to take a leak—maybe they
weren't on the fucking plane,' and you go through this kind
of denial syndrome . . . Maybe they'll find them caught in a
pine tree. I mean you can't help it. But there wasn't a finality
to it. But it was important to have the service 'cause it drugged
you. Getting into a week's time we needed a finality type
thing, to say, 'So this is over,' for the kids, for the younger
ones particularly. 'Let's get on.' . . . And even after they found
the remains you go through a denial. 'Maybe it wasn't really
them. If it took this long to identify them, maybe they
missed it.' "

We are on the Indianapolis Beltway. It's late and the six-
pack is finished. Jerry sits back and closes his eyes.

"It wasn't pilot error. I don't blame the pilot at all. In
fact, I'd like to send a letter to his family. My animosity is
toward the FAA. It's not very direct, but when we were . . .
It was right after the miscarriage that they were going through
those hearings and the whole thing seemed to be a matter of
semantics at that point, about what the guy meant when he
said 'You're clear for a landing,' and my thought was and is,
'Well, Jesus Christ. They've been flying for what, thirty-five

years, and they're talking about terminology? What in the fuck could they have been doing?' "

LT. JAMES E. ("MIKE") BROWN, 22, BODY BAG #143

Arlington National Cemetery, where Lieutenant Brown is buried. Standing at the information desk in the visitors' center an old woman asks about yet another dead soldier.

"I'm looking for the grave of the son of a friend of mine," the old woman says.

The girl behind the desk is expert, polite. "Yes . . ."

Unprompted, the woman provides details. "A captain in the air force. He came back from Vietnam badly wounded, but insisted on going back. He was shot down."

There are an average of three hundred funerals a month at Arlington National Cemetery.

"What was his name, ma'am?"

"Huh? . . . Bruce Johnson."

The girl goes through her files.

Her thin print dress blowing in the October wind, the old woman looks frail. "You can't live your life for them," she says to no one in particular.

The girl comes back. "I'm very sorry, but Bruce Johnson doesn't exist."

Lieutenant James E. Brown, however, is in section 39, just south of L'Enfant Drive. "Buried with full honors," the girl says. His gray-white marble government-issue marker is in a new row, down where the cemetery flattens out toward the Potomac, a row of yet unfaded stones in a place for old men filled with the young.

Mike Brown's mother wrote to me after she heard from Bettye Pierce (the mother of Dave Pierce; the two boys had

flown to and from Indianapolis together over Thanksgiving)
about this book. I replied, and she responded with a second
letter:

Fairfield, California
June 1, 1976

Dear Adam (if I may),

I appreciated your letter and fully intended to answer in
detail; however tragedy—in the form of my husband's death
on 1 May struck us again, and I've been literally reeling from
shock.

My only daughter's wedding was scheduled for 8 May and
thanks to many wonderful friends it went beautifully. Cheri
had said earlier [that] the only thing to mar her complete
happiness was the fact that "big brother"—Mike—wouldn't
be there; then of course Dad left us too.

My response, briefly, to some of your queries: feelings and
reactions to the crash: utter numb disbelief—then holding on
to hope of Mike's having possibly survived. Then the agoniz-
ing days that followed while waiting for positive identification
to be made. What a horrible nightmare, imagining all kinds
of disfigurement, suffering, etc. As in the case of my husband
(or so I'm told anyway) that death was instantaneous. I thank
a merciful God.

My husband held both private and commercial pilot's
licenses AND was (as a hobby) both an A/C mechanic and
a licensed FAA inspector, so was well-versed in flight procedures.

In bringing suit [against TWA] we hoped for a number
of things to be corrected to make flying safer for those of us
who log any kind of flying time for any purpose. We hoped to
bring pressure to bear to correct apathy on the part of pilots,
crews, ground controllers, etc.; to bring about changes within
the FAA to clarify responsibilities and semantics as related to
a uniform meaning and understanding nationwide. Further,
that the ground control device which was supposed to have
been mandatory by December 1, 1975, be installed in all air-

liners. It is doubtful that even now many airlines have placed them in their aircraft (claim too expensive. How about value of human lives???). Do know that many airlines have gained postponement approval!

Am so glad we did not go to trial in the end. Understand it was a mockery to those who lost their lives in that negligent and so very UNNECESSARY crash, and sheer torture for the loved ones left behind. It would appear to me TWA and the FAA came out "smelling like a rose."

Mike was 22 years old, a beautiful young man who had ideals—believed in God, country and family. He had graduated from the University of California at Berkeley with honors —from school of criminology. Planned to fulfill a commitment to the Marine Corps and possibly get his Masters while so involved. He had given consideration to serving [wanted to get into investigative part of USMC] a period in the FBI, but ultimately to go into Criminal Justice. He was a home-loving individual and this ultimately caused his death. Forgive me— I cannot go on at this point.

(signed) Nina Fowlston.

LaVada and Ron Hair, Mike's aunt and uncle, whom he visited for the holiday, say, "Mike just didn't want to spend his first Thanksgiving away from home in a lonely place like Quantico."

MARY ELLEN BROGAN, 48, BODY BAG #118

The kids have gathered to wait for their father. Darlene, the eldest—"Red" they call her; Bob, a little shaggy, the ex-merchant marine; Virginia, with long braids; barrel-chested George; and Sue, skittish behind her dark eyes.

"Oh no, Dad doesn't live here anymore—he moved out. He lives with . . ."

"... Betsy. Another woman."

"Yeah."

"He can't tell shit from apple butter."

Laughter.

A car pulls up to the curb on Oakwood Road.

"There he is."

"Is he drunk?" The question rips through the room, asked in unison.

"No—doesn't look like ..."

Kevin Brogan walks in, hair still wet from a shower. He wears a pressed plaid shirt, dark work pants, shined black shoes, white socks, glasses. The embarrassment is palpable. "Hi, Dad ..."

He avoids their eyes, gingerly sits in a small, padded armchair, and outlines the life he shared with Mary Ellen Brogan: married in '49, five children in as many years, work—she for the government, he for IBM—buying a house, then a trailer down by a lake ... The kids listen as he talks.

"Every Friday night we'd go down there all summer long, the two of us. As the kids got bigger they didn't particularly care to go down"—glances are exchanged around the room—"and there wasn't much for them to do. I mean, there's fishing and stuff like that ..."

What do you catch?

The boys: "Mosquitoes, beer cans, colds."

The girls: "No. The lake's very clean ..."

"Anyway," Kevin continues, "she had a premonition something was going to happen—three incidents. We were down at the lake when we closed up the trailer last September or November, shut off the water and the electricity and we backed out in the car to go home and I looked at the place and I says, 'You know, we've got a lot of work to do down here this coming spring.' She looked at me and she says, 'I don't think I'll be with you in the spring.' I says, 'What the hell are you talking about?' She says, 'I don't think I'll be with you in the spring.' I says, 'Damn it, get in the car.' So, we did." Pause for breath. "Well then, Thanksgiving if you remember, she got right here in that doorway and she says, 'I think this'll be the last Thanksgiving dinner I'll fix in this house.'"

"Yeah," says Darlene, "even my grandmother said—"

"Even her grandmother said something to her," Kevin finishes.

Darlene: "'Cause my grandmother, my mother's mom . . . George would take her home on Sundays and junk like that, but on Thanksgiving when George took her to go home, Mom came out on the porch and kissed goodbye and said to George, 'Take good care of your grandmother 'cause you know, she's not always going to be able to get around.'"

Kevin: "Then the day we took her out to the airport, Darlene and I took her out and uh, we were standing there and I kissed her and she grabbed my hand and she says 'You know, I don't really want to go,' and I says, 'Aw, go on.' I says, 'You get on that plane with Beverly Small [another one of the Zanesville HEW employees on Flight 514] and you'll be all right. You probably just got some butterflies.'"

"And then the birds," says Darlene.

"It was a month to three weeks before she got killed," Kevin explains. "I was on the second shift and she called me at work and she says, 'Kevin, there's a bird in the house.' Well, she was upstairs in the bathroom . . . 'Course we got a little landing right here and the bird was sitting on top of the curtain rod, and I said, 'Well, all you have to do is to go downstairs and open the front door. The bird wants to get out of there as much as you want it out of there.'"

"You know that movie, *The Birds*?" Virginia asks.

Kevin: "She's seen that movie, see . . . But Susie, Susie wants to tell you about . . ."

Susie: ". . . the old wive's tale of a bird in the hand . . ."

The talk of premonition helps the conversation become general. Still, Kevin and his children are uneasy with each other, tentative, as though meeting for the first time after a fight and groping for common ground. When I leave, so does Kevin. We agree to talk again, just the two of us.

Green-eyed Virginia (with the long braids) calls two days later. It's raining hard outside—spring in Ohio—a good day for a movie. She suggests *Tommy*. We work out directions and logistics. Just before hanging up, almost as an after-thought, she says, "My father *never* wears white socks, never."
"What?"
"You know, the other day at the house?"
"Yeah."
"I've never seen him like that, I dunno . . ."

Virginia and I meet in a cavernous mall with plastic foun-tains and piped-in music. She's too thin and wears too much blue eye-shadow. Even her freckles look pale. The movie theater is huge, empty except for a cluster of love-beaded freaks who beat time to the blaring music and make frequent trips to the bathroom. After the picture Virginia suggests a drink at the Red Dog Grill down the road—one of those places hung with imitation Tiffany lamps and twenty-foot-long salad bars.
Virginia drinks sloe gin fizzes through a straw. Mom, she says, really died at the wrong time. Things were going great. Dad had quit drinking—she'd threatened to divorce him if he didn't—and they were like two lovebirds all over again. And Mom loved her work. She was at the beginning of a whole new career. She'd gotten a promotion down at HEW. She was a professional. The eighteen years of clerical work were over. She had responsibility, she went out to lunch with may-ors and politicians all over the state. And the two boys were in school and working, and Susie had a decent boyfriend for a change and . . . then they went and killed her. And Dad started drinking again and "then for an escape or to grab onto something, he picked up with her—Betsy."
How does the rest of the family feel about it?

"Everybody, you know ... She was our mom. What else could you say? We loved her."

No, I mean about this woman?

"I hate her."

Really?

"I don't know— The way he brought her in, like I said— if he'd brought her in like a father would—like when he brought her in he was drunk. I walked in from the kitchen to where we was all sitting [in the house on Oakwood Road]. He was sitting on that flowered couch, laying right up over it, I mean, just completely. He said, 'Aren't you gonna say hi to Betsy?' And I looked over and when I seen him I started shaking you know, and Bob came up and Red grabbed me and you know, Dad said, 'What the hell is wrong with her?' I said, 'Man, Dad, sit up!' That's the first time I'd ever seen my dad with another woman—God, it just hit me too fast and he was drunk ... After all the meals Mom cooked and shit like that ... We got into a big argument ... Then I went and seen this psychiatrist—I'd go high, I'd go high to the psychiatrist. I'd leave my job smoking a joint on the way to the psychiatrist, get in there, and we'd be talking and I'd laugh at him, you know? I went three times." (Kevin paid for the visits.) "It was thirty-one dollars and, shit, I took so many dope pills my hands broke out—they got scars on them. And my stomach started really fucking up again, I couldn't eat. I'm still ... Like when Mom got killed, I weighed one-twenty-six. I'm one-twelve now and I keep going down ... But, like, back to her. When it all happened, I stayed high and I've been high ever since—except like when I go to work."

Does it help?

"Yeah ... You know, my mom and I didn't talk for about a year, about two years ago. Me and her and Dad went through a hell of a thing. Red—Darlene—wrecked the family car and they felt I had to give mine up because she was a beautician [and needed it to get to work] and we got in a big fight about it. And like, I moved out and haven't went back since. But, you know, they'd come over to my apartment and shit like that. Then I was living with some guy and they didn't like it and so like all of a sudden there was nothing there, and I moved home for about two weeks until I got myself another apartment. My mother didn't want me to leave, but then she treated me as another woman instead of a little girl. Like she'd come over and I'd make pineapple upside-down cake and Mom would get into it—she was really hip but she was still old-fashioned and a mother. You never lost your respect."

Another sloe gin fizz—red soda pop.

"Like I wasn't angry [about the crash] until those men from TWA started really fuckin' with me. When I read that stuff in the paper about the busload of people who went up there, to the crash site, I called them. They got meaner than a son of a bitch and they said they could only authorize one person to go. [She wanted to go to Virginia.] And they wouldn't pay for a hotel or nothing! And I really laid them out . . . I want them to *pay*!"

And now?

"Now I'm getting used to it. I missed her at first like all the time, but even now it's still hard to believe she's gone . . . You know, she'd call and we'd talk and we'd go shopping— the little things is what you miss. Just going over there and her being in the whole house completely. The house is empty now. I mean, everybody could be there—it's empty. You have the feeling like you're waiting for her to walk out in the kitchen or down the steps and she's not. But I've accepted it. It's just that, I don't know, like I don't feel the real feeling of it, you understand? I don't know whether it's because I didn't see her laying in her casket—it was closed, you know, during the funeral. The priest and all this shit was there, I was kneeling down, I tried to visualize what was in the casket. I didn't know if it was . . . I wondered was she just in there, in a bag? Those plastic bags. What was she like? That's freaky."

Her face is now that of a little girl, wide-eyed, puzzled,

afraid even of the words she's saying. She takes a long swallow from her drink, collects herself, and slips back into the tough talk.

"The funeral didn't do shit. Because it was like someone else's. This shit the priest tried to tell me—'Virginia, her soul's gone; that's just a body, it's really nothing.' Well, that *was* her, that's the body, that's the thing I knew."

Bob Brogan has a lot on his mind. Some of it has to do with the general disorder of things, of trying to fashion a life after three lackadaisical years in the Merchant Marine, and some of it has to do with his mother's death—his mother who had nudged him to go get a job, to apply to college. His mother who is gone now and his father who, he says, in any real sense, is also gone. So Tom, sitting in the weak sunshine on the porch of the family house, sips at a beer and tries to work it out.

"When I came home [at age twenty-two] it was different, 'cause I didn't want to do anything and they wanted me to get a job and everything. Oh, I came out with about a couple of thousand and flashed it around . . ."

George, the younger brother who is listening in, can't resist. "Yeah, he thought he was a big shit or something."

Unperturbed, Bob continues. "Then I went down and signed up for unemployment, I wasn't gonna work, I was gonna take the whole unemployment, and so I bummed around . . . and after two months of being unemployed, they started hassling me—Mom, really. She used to get on me about my hair, about getting a job, getting into school. Well, after my unemployment was starting to run out I decided to go look for a job, and a dude hired me for some reason. I was a building supervisor at a high-rise downtown," which is where he was the day of the crash.

"No one knows this, but I sat at work and cried for half an hour . . . 'Cause I went to work that morning—I worked on Sundays. I worked six days a week. Mom was up and was yelling at me 'cause I was late, saying, 'You're gonna lose your goddamned job,' and all this and that and I said, 'Well, I don't care, I'm gonna quit in a couple of weeks anyhow, to go to school,' and she said—she was yelling—she said something, and I said, 'Aw, go to Washington and I'll see you when you get back.'"

He shrugs, wishing their parting had been different, shrugs knowing there's no changing it. He helped carry her casket at the funeral, a filial pallbearer. Struggling to find feelings which he thinks ought to be there he says, "I look back on it and at least I did do something."

For the first few months, in early winter, the Brogans would troop to Sunday Mass together and, says Bob, "In some ways, it brought us together, the family was a little bit better, Dad was still not drinking . . ."

But that, apparently, did not last. Kevin moved out of the house, the boys went to school, the girls to work, life settled into an uneasy routine in expectation of the settlement from TWA. By accident they all met, one muddy afternoon down at "the lake."

Bob gets more beer, qualifies the whole incident as "a hassle and a half," and settles down to explain what happened.

"Me and Darlene and Virginia drove down with the three dogs. We arrived and started cleaning up. I got the hand mower out and started cutting, and around noon Dad pulled up with Betsy."

From the floor of the porch where he's sitting, George mumbles, "I knew it and I stayed home. I didn't want no part of it."

"It was just two weeks ago, three weeks ago. It'll be three weeks ago this week," says Bob. "Anyway they pulled up, Betsy and the three kids [hers from a previous marriage], the tractor and bicycle and a couple of toys for the kids. So I go ahead, I start unloading the tractor [from Kevin's green pickup truck] and we're going to drive it off and then I changed my mind so . . . Meantime the kids start yelling about 'I want my bicycle, hurry and get that tractor off,' so I picked up the bicycle and kind of threw it out in the yard, and, uh,

I pushed the tractor down and started it up and started cutting grass. So the old man goes out and unloads the truck and takes Betsy up on the porch and she takes the baby and he goes in and gets a beer and says hi to Darlene and all this. So I finished cutting—ran out of gas—so I stopped the mower, and Virginia pulls up.

"I said, 'Dad, the tractor's all yours; you go ahead and run it—just put some gas in it.' So Virginia gets in there and says hi and walks in and comes back out. And started in on him about not doing anything—he'd just got there—you know about 'why don't you get off your ass and go cut the grass' and all this. I wasn't really paying too much attention—we were fixing the plumbing—and, I don't know, Virginia got onto him and he ain't doing anything, went in for another beer, and he come back out and they were arguing about Betsy, [Virginia saying] 'You can just take her the hell out of here.' She says, 'I don't want her down here' and this and that, but the old man started beatin' on her; I heard him hittin' her . . . inside, in the bedroom. So I ran in the trailer and I just grabbed him and threw him against the wall. I said, 'Don't hit her,' I said, 'we can talk sensibly and argue sensibly . . . you don't have to hit her—she's your daughter, you know, I mean you don't have to beat her up just 'cause she can express herself.' And so he went out and gets the gun— a .38 special—and comes in and I'm standing in the kitchen with Virginia and Darlene, and he comes in and says, 'Well, I'm gonna shoot your ass.' So he puts it to my head. And so I said, 'Go ahead. But you won't.' I said, 'I'll be laying next to Mom and you'll be in the pen somewhere.' And he says, 'No you won't. Anybody that's gonna lay next to Mom, it's me.' I said, 'When you're in the pen for years and I'm buried next to Mom and we're *watching* you . . .' He said, 'The hell with you,' so I shook the gun away, and he put it in the holster and he said, 'You're lucky I don't kill you for keeping me away from Virginia.' I said, 'Virginia, get out of the kitchen. Dad's drunk. Go sit on the porch.' "

Eventually, Bob says, they took Kevin's gun and emptied

it and hid it. Realizing that he was not going to get his revolver back, Kevin apparently packed Betsy and her children
in his truck, had one more beer, and roared off.

His placid midwestern soul bewildered by his family's
eruptions of Irish temper, Bob chronicles the whole distaste-
ful episode in a listless drawl, accompanied by much head
shaking.

"Maybe I shouldn't a hit him," he concludes dispassion-
ately, "but, I mean, if you could have seen him sittin' there
going like this—" imitates a groggy prize-fighter roundhous-
ing—"full fist, kickin' her off the ground... I just couldn't
take it. I was gonna try and hold him, but I figured, well, I
might as well smack him."

And there was one other time, Bob adds in his "oh-shit,
why-do-I-have-to-get-involved-in-this" tone of voice, "one incident when he was so drunk... (he always picked on Virginia;
she had a way of getting to him)... he put her through the
upstairs window at the landing, fighting on Christmas Eve.
Mom was alive—I don't know... I don't know if Mom was
in on Virginia's case or not. I know I wrassled Dad into the
Christmas tree and tore that down. Then, he threatened to
kill himself. Mom called the cops and they came and took
him out, took him downtown for the night. That was the only
other incident."

But now there's the question of money—rather, the expectation of money: the settlement from TWA. The money,
Bob says, would help. "I mean it'd give me the things I'd
probably have to work hard to get, like it'd set me up if I
wanted to buy a house or something like that. I'm sure I'll
get that much. But price-wise if I don't think it's fair, I'll just
say it. Even the lawyer said since there's so much conflict in
the family, the next time he gets an offer, he ain't gonna invite just Dad in to talk to him—he's gonna invite the whole
clan in."

But actual money, however, has already come in and
caused dissension. Mary Ellen Brogan carried government em-

ployee's and labor union insurance policies which, says Bob, netted Kevin close to $60,000.

The night before Memorial Day Kevin, who by then had already moved out of the house on Oakwood Road, returned. He wanted his gun back, the .38 revolver Bob had taken from him down at the lake.

Bob told him he'd thrown it away. His father told him to go and retrieve it. Bob blankly turned to him and said, "I know what you're gonna do. You're gonna get us all here, fake the money"—Kevin had said he'd give each child $1000 the next evening—"and just start shooting when you come in the door and then just turn the gun on yourself."

"I can always go out and buy another one," Kevin told him.

"Well, goddamnit, go out and buy another one then," Bob fired back, "but you step in that doorway with a loaded .38 and that's the way you're goin' out—with a slug in your ass."

"What do you want?" Kevin demanded. "You want a showdown here tomorrow night?"

"Go ahead," Bob said. "I'll be waiting."

Kevin returned at dusk on Memorial Day to give everyone but Bob the promised $1000 in cash.

Bob asked for his share. His father said, "Give me the gun."

Bob told him he'd emptied the .38 and thrown the shells in the lake, "so you wouldn't shoot nobody."

The gun, said Kevin.

"Betsy's got it," his son told him. "Sue and Virginia took it over."

Kevin called Betsy. She confirmed she had it.

"I'm taking off a hundred dollars for every bullet gone," Kevin said.

They argued. The other kids watched. Finally Kevin gave his son the thousand bucks. Bob thanked him. Kevin left.

Bob remembers "being kind of scared after he gave me the thousand dollars, 'cause he got his gun back. I figured he'd go out to his car, get his shells, and then come back or something. I went upstairs and was reading homework—I already had a couple of slugs in my shotgun laying beside me."

And the money?

"I figure he's buying his way out so I'm gonna let him go."

"I may cry, you know—I'm the big softie of the family." Darlene smiles, apologizing in advance for the tears to come. She knows the secrets and has reflected upon them—the shaky foundations, cracked walls, crumbling bricks, and dissolving matrix of a family structure which she alone now struggles to shore against apathy and the insidious rot of greed; greed awakened by the expectation of money and the resulting illusions of freedom and the easy life. Blood money.

To brighten her life she qualifies anything remotely pleasant as "super," takes pleasure in cheap Friday night suppers with the girls from the beauty shop, finds strength in paying the bills and feeding her brothers. Unwilling to settle into marriage as too easy an out, she waits for "Mister Right." Though no doubt envious of her two younger, prettier sisters, she would not be like them. No shiftless, violent boys, no drugs; rather—a friend, a good clean cop, a man who would call her "honey" or some little intimate nickname of their own. Darlene is twenty-six, irrevocably overweight, the "mama" of the family.

"At the funeral Dad was saying, 'Well, this is the little mother, she's gonna take over now,' referring to me, and that's what I did for a while—I took care of 'em because he didn't even know what was paid on anything, because Mom took care of all that. He turned everything over to me except for the administration of the estate. It was even my job, the day the guys from TWA called saying that they had my mother's wedding ring, it was my job to go and get it," over at Port Columbus Airport. "We got it and got two receipts that she had in her purse and forty-six dollars wrapped up in one recepit, and the Mother's Day ring I gave her and a necklace that Bob brought her back from overseas. I got a piece of the chain that was underneath her shirt collar. That's all there was of that. There was nothing to get back anymore . . . She was the whole backbone of the family."

And you're becoming that, aren't you?

"Yeah. I will, for the kids' sake. But for my Dad, no. He's as bull-headed as Virginia is, but there's no use of getting upset or yelling . . . Like the marriage my Mom and Dad had, my Mom put up with a hell of a lot, but she loved my father so. Ninety per cent of their arguments stemmed from his drinking. From what his brothers and sisters say, it's like a family curse on his family, for the guys . . . They come from a very small town in West Virginia. There was nothing to do unless you worked in the mines, or left town and found another job someplace else, but at night it was just sitting around drinking, and that's what he did, you know, until he finally left and came out here. By then his social pattern of drinking was there and it got progressively worse." There were arguments in the night, she says, her mother and father downstairs, the children above, awake in their beds, silently listening, "Wondering . . ."

In the later years Darlene says, "it got to the point where Dad was jealous of Mom and her job. See, as it stands today, my Dad is at the top that he can ever achieve to do; he can't go anyplace else." Mary Ellen had just been promoted at HEW, "and the fact for my Mom being forty-eight years old, she was a damn good-looking woman—for having five kids— the fact that her having to travel with, you know, different men and stuff like that, was bothering him quite a bit. I think if he hadn't quit his drinking last July for the period that he did, my Mom would have divorced him. She had talked about that—I think it entered her mind a lot, but like she came from a broken home, her father had a drinking problem, she was raised in that atmosphere and she didn't want us kids to be raised in the same atmosphere."

Spooked by the finally uttered threat of divorce, Kevin went on the wagon. "He was proud of those six months of sobriety; he was super proud of her and they had a super marriage then—they found each other again. It was a super good thing. Then, the day we buried my mother, he got drunk that night."

And later, after the incident at the lake, he fled the house and the accusing eyes of his children.

On a gray, lowering day in June, with the city's tornado-watch sirens shrieking outside, Kevin Brogan offers up his story. We meet, at his suggestion, on the southeastern outskirts of Zanesville in a small tract house he shares with Betsy and her two children. Supportively, she remains in the toy-littered living room while we talk, offering coffee and soulful looks.

Without prompting, he begins by talking about the settlement he expects from TWA. His lawyer "knows what's going on, oh yeah, and I guess Virginia called him one night and raised a bit of hell—Darlene called him too and raised a little bit of hell, feeling he's not acting fast enough or not doing anything. They did it unbeknownst to me. But, he says, 'I want to deal directly with you,' and I says, 'That's the way I want it and that's the way it is from now on.' He just absolutely will refuse to talk to them. I'm the one that's gonna make the decisions. He says it's gonna come down to the point where there's going to be some agreement made as to how much to settle on each child. 'And,' he says, 'you're the one that's gonna have to make the decision.' He has seen so many families broken up over money, he certainly wouldn't want to see it happen to my family. Well, I told him at the time, as far as I was concerned, what was gotten out of TWA —see, there's two lawyers handling this for me—I wanted it all to go to the children because I felt that I got enough out of the insurance and what I'm getting from the government— that'll take care of me.

"Well, they both agreed [that was] bullshit. You lost something too [they said] and you're entitled to something. Then the lawyer found out we didn't have a will, so the first thing he says is 'I'm gonna make out a will for you because,' he says, 'you can get out there on the freeway and get killed and then we got a lot more legal problems.' So that's the first thing he did. Then I set everything up that I wanted divided, equally among all the five children, regardless. Nobody gets a penny more or less than the other. So then we get to talking, the lawyer, and he says, 'I've seen it happen before,' he says; 'it's gonna break up your family and,' he says, 'I'd hate like hell to see that happen. I can set up a trust fund, X-number of dollars for each child,' and I says, 'all right.' So we let it go at that because nobody really knows what the settlement's gonna be. As far as Bob is concerned, he needs to go to school, he'll get what he needs. So will George. Uh, it'll help Susie to buy a house, she'll get a certain amount to buy a house. But I want a certain amount for a trust fund. So they naturally knew I had gone to see the lawyer and I told them exactly what I just told you, which is, I wanted to set up a trust fund, but they would get a certain amount of cash—if they needed it. Well, I got reactions from three of them. They didn't want a trust fund. Virginia was one, Susie would be another, naturally Bob was the other one. Now, they can get a lawyer. Well, I called my lawyer the next day, after I got the reactions from them, and told him the reaction I got. I didn't get any calls from Darlene or George. George is a saver anyway—I don't get no problems from George. I told the lawyer what had gone on and he said, 'Well, they're not under age; they can really fight it if they want to, and there isn't a goddamned thing [you can do] except fight them back—then it'll be up to the courts.' So that's the way things is set up for the present time."

Kevin sits back, switches from coffee to beer.

"There was a little money problem come up here possibly two weeks ago and, like I says, I just said, 'The hell with it . . . I'll give you each one of you a thousand dollars and that's gonna have to tide you over. I don't give a good goddamn!' Now, Bobby didn't need it because he's getting help from the government [a GI bill scholarship and living allowance]. George didn't need it to go to school because he's also get-

ting help, from Mary Ellen's estate [the civil servant's life
insurance paying for schooling for minor children]—and he
works. Darlene didn't really need it. Susie could use it, and
Virginia could use it. And this is where I got most of the
static from. So, like I told 'em, I'd just go down and I'll get
five thousand dollars and give you all a thousand dollars
apiece. Then, boom, the hell with you. That's it. Just don't
bug me no more.

"I mean, how long is the old man going to come along
and say, 'Well here now—here's a hundred bucks, here's two-
forty, here's a thousand.' I mean, after all! You keep getting
them out [of a bind] and they're gonna keep getting in. That's
like I told them when I gave them money the other day—
'Now look, babes, this is it. You just got to make it on your
own'—and they can make it on their own. There's no reason
why they couldn't. Before this happened we were making a
good living; Mary Ellen and I were both working and making
fairly good money, we were living comfortable. But we were
not in a bind. But since it happened, they seem to think you
know, you've got this money now, I can go out there and get
in here and Dad's gonna help me. And that's the way it's
been since this happened.

"Besides giving them five thousand that I just gave them
last week, this February they wanted five thousand other dollars, that I gave them. So that's ten thousand in less than six
months. So there ain't no goddamn way...I mean, that's
the end of it. Now, it might sound bitter, it may sound wrong,
but if I do it, then the first thing it's gonna be fifteen thousand. And just...there's no way I'm gonna do it. I laid down
the law to each and every one of them the other night when
I was up there. There ain't no goddamn way I'm gonna do it
anymore. You can take this and put it in the bank and save
it; if you get in a bind, go ahead and use it. So if you want
to go out there and throw it away, that's your problem, not
mine. Don't come to me no more." The righteousness in his
voice trails off. "Of course, if they came to me and said they

were in trouble as far as sickness or something like that, that's different."

But you moved out of the house?

"Yes," Kevin says, "yes." It happened after the flare-up down at the lake. He came back to Zanesville, packed his bags and moved in with Betsy—"I rent the basement."

The lake?

"Well," he says with a tired shrug, "we got into a family argument, name-calling, things came up that shouldn't have come up—about their mother and stuff. They called her—" he nods towards Betsy—"a you-know-what—stuff like that. Very unnecessary."

For the first time, Betsy speaks up. "I think in a way it sort of brought the kids closer together. I think maybe they'll feel closer to him now, uh, maybe they got a lot of stuff off their chests towards me and, uh, I think maybe . . ."

"See," Kevin says, "they feel . . ."

"It's a funny thing, but it could have been a good thing to happen," Betsy insists in her woodsy Kentucky accent.

"What they're afraid of," Kevin goes on, "and what my sisters and them are afraid of, and even my mother's afraid of is that, bang!—I'm gonna get married right now. Possibly you've been thinking that yourself. I don't know. Maybe you was gonna ask me that. Well, it's like I told Betsy—there's no way I'm gonna get married right now. No way."

Betsy looks away, feigns disinterest.

"I think they resented the fact that I met Betsy, I really know they did. Really." Kevin stops—brusquely changes the subject. "Hey, I go to a psychiatrist and I go to Dr. Armstrong—he's the family doctor—Darlene goes to him and Virginia goes to him, and I go to him and to Dr. Hayes, the psychiatrist. I'm the only one—Virginia was going to her own psychiatrist. In fact, I just paid her doctor's bill last week. So, uh, see, there's a period of about three months when something like this happens, according to Dr. Hayes and Dr. Armstrong—it doesn't hit you, only subconsciously, and then it hits you and this happened to me. And to Virginia and I think some of it happened to Darlene. And, uh, that's when I started going to a psychiatrist." Kevin sounds both bewildered and grateful that psychiatry—a form of help formerly beyond his ken—has helped.

"I just damn near went out of my mind," he says earnestly. "I couldn't sleep, I wasn't right, I was drinking. I didn't give a goddamn for anything—believe me—including work. And I've worked for this place [IBM] for twenty years and haven't missed a day. But, [after the crash] to hell with everything is the way you felt. You felt you wished to Christ you could die. And then I started spending time in bars, uh . . . sit and brood, wouldn't eat right, couldn't sleep right and finally I did go to Dr. Armstrong," who put him on medication and recommended that he take some time off. Kevin drove to Florida and "when I come back I could find myself talking more and more about it, I could talk about it fairly freely, and it don't bother me too much as it used to."

What would cross your mind when you did think about it?

"How it happened, why it happened. And what actually she might have thought at the time it happened. I could see her torn to bits . . . uh, Mary Ellen was a person who didn't like to be hurt. I mean, she couldn't take hurt and I've seen her be sick but she wouldn't admit being sick. Uh, then I got the death certificate and on there it says 'serious lacerations' or something like that was the cause of death ["multiple mass injuries" read all 92 death certificates] and I couldn't stand this. I mean, to me, she was just torn to hell. And knowing her and knowing she didn't want to be hurt, I wonder what the hell she thought of. But like Dr. Armstrong and Dr. Hayes both told me, it happened so goddamned fast that they didn't even have a chance to think about it." He considers that fact, hopes it is true. Silence.

How long were you married?

"Going on twenty-six years. Like everybody else, we've had problems with the children—I don't mean problems in the sense connected with the police or connected with dope or connected with . . . They do certain things we didn't like. We'd talk about it and try to advise them, uh—Mary Ellen didn't like what Susie did, she didn't like what Virginia did— like she cried over it maybe at night. But I told her, 'It's their life and they're young adults. There's not a goddamn thing you can do about it.' But she didn't like it. It broke her heart, I'll tell you the truth."

You mean because of their moving out of the house and the kinds of lives they lead?

"Right, right, right. I mean like I says, they're young adults. We tried to bring them up right. We never had any problems as far as dope or connections with the police . . ." He looks at me, wondering perhaps how blunt Virginia had been about herself.

Would you consider going back home?

"Yes, uh, we've talked about it . . . Possibly it would be better if I went home and had more control over it, till something happened, till I retired or bought a farm or something like that—things are going to depend on whatever comes out of the settlement and if I get a hold of this farm I've been thinking about, I hope to get the hell out of IBM before I'm fifty-five"—in three years. And, "Bobby's talking about going to California as soon as he gets out of school; George is hoping he gets into the highway patrol, so actually, the only one's gonna be home is Darlene, and that big barn [of a house] is a little too much for her. So, all in all, it probably would be better if I did go home."

He thinks about what he's just said. His eyes wander over to Betsy, who looks away. "Uh, you don't want to do what's wrong, naturally," he finally says.

Of course . . . The other night I remember Virginia said that the last nine months of your life together with Mary Ellen were like a second honeymoon.

"Well," Kevin begins, "I always drank beer—I never drank liquor—I drank before I even knew Mary Ellen, and she didn't like my drinking because her Dad was an alcoholic. He died an alcoholic and she grew up with it and she didn't like it. And she's got one brother that's an alcoholic right now, he's in and out of the... He's the one that spent twenty-two years in the navy. But, uh, Mike's like his Dad—it's a sickness, it's a disease and actually, right now, people are starting to recognize that. And the Lord knows, you know as well as I do, it's a dope, it's a drug. I don't know if you ever heard of *anibuse* [antabuse]. Well, we got talking one time and she wanted me to quit drinking. We went to Dr. Armstrong and talked to him, and he says there's a drug on the market that will make you quit drinking or you'll get sicker than a dog. Well, I didn't touch a drink from June until after it happened. I felt beautiful. I felt the best that I'd felt in years, for not drinking. And," he says emphatically, "my relations with Mary Ellen were, hell, ninety-five per cent better. The kids were happier, and my sisters and my mother, they couldn't hardly believe it. Everything was a hell of a lot better."

But, after the crash, to keep the nightmares and the shakes at bay, he started again. "In fact," Kevin says, "Dr. Armstrong tried to get me back on anibuse. I've got them pills out there but, uh, I still stop and have a beer after work. I don't drink as much as I used to... One or two glasses of beer and then I'll have three or four cans here in the evening, but that's it."

Betsy nods in agreement.

"But hell, I used to sit down and drink a case of beer and think nothing of it. Weekends—two cases..."

Mary Ellen... did she... were there any hopes, expectations in her life she either had met or hoped to meet?

"Well, she tried to get as high in HEW as she could go. I don't know how high she would have went with them, of course, nobody knows, but, uh, they all liked her. I think she was well satisfied of what she had achieved because she came from... Well, they have two branches there, your clerical and your professional. And she had just made the jump, and to her, this was wonderful, believe me. She tried for that for years—she went to school, she studied—Christ, she'd take work home, those manuals and stuff, my God! Because she was still a young woman really, Mary Ellen was smart, uh, she still wanted to continue on with her schooling, like if something new came up, something interesting, she'd put in for it. I think that's what happened in this goddamn school" —meaning her training trip to Washington on Flight 514.

"I wasn't particularly crazy about her putting in for school. See, she'd gone the year before for a month and she'd been promised certain things and she didn't get them. They didn't realize or they didn't materialize. So she was disappointed and so was I. So when this chance for schooling came up I told her, I says, 'It'll probably be a bunch of bullshit like the last one was, and you'll wind up with nothing out of it.' Anyway, I wasn't that happy about her going but, uh, like I say, she wanted to go so I didn't object. I never objected to her if she wanted to go to school or if she felt she wanted to do this or that."

And you?

"Years ago I wanted to buy a bar, but she wouldn't go along with me."

ELIZABETH WRIGHT, 19, BODY BAG #8

"Then I walked her to the check-in point, gave her five dollars for cab fare, told her I loved her and was proud of her. I gave her a big hug and kiss and told her to write home more often. The last thing she said to me was, 'You too. You're not any better than I am.'"

On a crisp November afternoon, Jo and Bob Wright gather around a coffee table spread with yearbooks and photographs. They talk about their daughter, laugh at some memories, unashamed, cry at others. The wind blows dry twigs against the window panes: football weather with the smell of winter coming on.

A shy, quiet, short girl with a "mousy little voice," Liz

blossomed late, they say, spurred by a passion for acting, rediscovered on a college gymnasium stage. Whatever its intellectual shortcomings, Marymount Junior College in Arlington, Virginia ("It seems girls' schools are not hard to get into any more," Mrs. Wright says, trying for a smile; "they're so expensive they'll take you if you can walk—Catholic girls' schools in particular") was fascinating to Liz. In a letter home to Lancaster, dated "The first day of classes," she wrote that she was the only freshman from Ohio, a fact only slightly less "amazing" than her compatibility with her roommate who "promised to fix me up with a boyfriend of hers, so everything is A-OK."

Enticing as the ideas of boys and nights in Washington may have been (the nuns at Marymount apparently understood such things—their advice to new girls being a mild admonition not to go to a young man's apartment on the first date), Liz found acting safer and more exciting. And while her rich classmates practiced their lines on the Georgetown boys across the Potomac, Liz euphorically wrote her parents that Boyd Hagy, her drama coach, "promised to help me to anything to get started in the theater or anything to keep me out of it (if I stink)!"

Recognizing in Liz an uncommon candor ("She was not a raving beauty and knew she would never be a leading lady") coupled with equal determination ("She was one of those few young girls willing to play character roles"), Hagy cast her as Mrs. Bramson in Emlyn Williams' 1937 British murder mystery, *Night Must Fall*, Marymount's 1973 fall production. The stage directions called for "a fussy, discontented, common woman of fifty-five, old-fashioned in both clothes and coiffure," and Liz's debut on stage consisted of hurling complaints and vilifications at her ficticious family while confined to a wheelchair.

Though Mrs. Bramson was smothered to death by a pillow-wielding psychopath at the beginning of Act Three, Liz pedaled back to life for Marymount's spring production as Madame Arcati, Noel Coward's bicycle-riding medium in

Blithe Spirit. Called this time to portray "a striking woman, dressed not too extravagantly but with a decided bias toward the barbaric," Liz strode the gymnasium stage munching cucumber sandwiches (real) and drinking dry martinis (not real) in an enthusiastic performance which established her as a star among the college's eight drama students.

Elated, she returned home for the summer steadfast in her resolve to become an actress, yet willing to satisfy her mother's plea that she study something a trifle more "secure" than drama by a good-natured promise to take up typing when school resumed.

The Wrights are fond of the story about how Bob's brother studied to be an opera singer, in Europe, for seven years. "And now," says Mrs. Wright, "he sings in the choir with the Atlanta Glee Club. So, there we are; we've got a son who is an artist, Lizzy was the actress, and Maggie [the youngest] likes to write."

Jo used to be an airline stewardess, and Bob is an executive for the Anchor Hocking Corporation, which makes glassware.

It is suggested that Liz might have taken, along with typing, a course on waiting on tables, the age-old unemployed actor's meal ticket.

"Right. Absolutely," says Jo. "And she did, I mean she worked at the Burger Chef the last summer. And at night, when she was off, she and her boyfriend were in a one-act play out here. She'd serve those damn hamburgers all day . . . We told the kids we'd pay for their tuition [nearly $3000 a year at Marymount] but as far as expenses, coming home, other than Christmas and things like that, they'd have to make do on their own. So, she goes out and gets a job and makes eight hundred dollars; she had it figured out, four hundred for the first half of the year and four hundred for the second."

With Washington cooling after the long summer of Watergate and a presidential resignation, Liz returned to Marymount and auditioned for and received the lead in the fall production, playing Catherine in *The Heiress*—a bland girl who, as the play progresses, is forced by a lover's betrayal and an egotistical father's intransigence to fathom her own identity. Catherine emerges at the end bloodied, alone, but her own woman.

As the gentle Virginia autumn wore on, Liz plowed into her school work and her new role, memorizing her lines sometimes by mumbling about her room, at times cued by Susan Tarasi, a volatile, raven-haired beauty with whom she vied for parts and recognition. With the play in rehearsal and examinations in the offing, Parents' Weekend late in October proved a welcome break.

On a Saturday night a week later, Liz wrote home:

Dear Family,

Well well well—it has been quite a Parents' Weekend being adopted by the Tarasis and going out to dinner Friday night at Clyde's [the P. J. Clarke's of Washington's Camelot years] at Trader Vic's on Saturday in Wash. D.C. [Nixon's old favorite] and Sunday night at Gusti's Italian restaurant [the suburban crowd's favorite]. As a result it was quite some weekend—at least a 5 lb. gainer. Ha!—well I guess it was truly worth it because I had a blast and I dieted all week and I'll probably never get the chance of dining like that again.... I so wished you Mom and Dad and Marg could have been here to share the weekend with me—I really missed you—you could have been at the Honors Award ceremony because I did achieve Honors membership. I feel extremely proud of myself for many reasons: 1) because I never thought I could ever achieve anything at all—I've always felt so inferior and average as far as academics go; 2) it's totally made me realize I can do anything I want if I want it badly enough—all I have to do is truly try and it will come. The reward is so fulfilling— knowing that you worked so hard for something and proved it to yourself that you could—oh, it's hard to explain but all I know is that I'm very happy with myself as far as academics go—well that's just a small reason why I wish you were all

here—mainly because of a problem I have right now—it seems
I'm having a great deal of trouble making friends, ya know—it
just seems that I'm not too well liked this year—maybe I have
changed or something—but people aren't as friendly this year
—they're extremely snobby and self-centered, living in their
own little worlds of security—I guess we all do that to an
extent—but whatever happened to the good 'ole days of going
out and meeting "new" friends?? People change, but I can't
accept it—it makes me have so very little trust in kids here.

Forget it as far as dates go—I've truly tried—and I can
see. . . . .

The letter ends, unfinished, unsent. Sophomore year au-
tumnal depression. Dateless Saturday nights. Doubts. Growing up. Nothing a standing ovation blushingly received on
stage could not and did not cure.

The Heiress played for two nights in November, just
before the Thanksgiving break. Hailed by a judge for the
American College Theater Festival as the finest small college
production he'd seen in twenty years, the play was nominated
for production in Washington's John F. Kennedy Center
for the Performing Arts. It was hinted to Liz that next year
she would get a drama scholarship at the University of North
Carolina and she flew home for the holiday skinny, self-
assured, a success.

"It's so beautiful to see these girls come home from college and see how they change," her mother says. "It's beautiful."

The Wright's son had just gotten married; Thanksgiving
was a festive affair full of promises, love, and snow.

Then came Sunday—and the rest.

"I felt I was in a glass dome," Mrs. Wright says. "It hurt
all over, it was hard to walk, I had to hold onto the walls and
the furniture. I could only whisper. For the next months I
tried to stop thinking, tried 'keeping busy'—painted bed-
rooms, upholstered chairs, made slipcovers, more clothes than

Donna [her new daughter-in-law] could ever wear, made shirts for Maggie to fill two closets, clothes for me that if I live to be a hundred I could never have time to wear."

TWA's three-paragraph form letter of condolences arrived ten days after the crash. Signed by F. C. Wiser, the airline's president, it informed them, the Wrights said, that though Wiser couldn't hazard fathoming the grief they felt, and though there wasn't much he could say to ease their grief, he hoped it would be of some help to know that TWA shared it with them.

Infuriated—"I hate the man who wrote that letter"—Jo Wright wrote back insisting that she and her husband receive a personal apology. They did, settled their case against the airline without going to court, and used some of the money to set up a Marymount scholarship in their daughter's name.

"My only regret is that the damn huge tuition of two kids in college did not allow us to see Liz's triumphs," says Jo Wright. "It's a bit of heaven to see their dreams come true."

Relentlessly tormented by the senselessness, the wantonness of Liz's death, Jo, after exhausting the comforts of her religion, briefly committed herself to psychiatric care. But her nightmares persisted and she began jotting down what she remembered of their horror. This is what she wrote on the morning of June 16, half a year after that Thanksgiving Sunday:

We were at Marymount—Liz asked where was everyone. Many teachers in rooms (I knew it was summer school). Some local nun surprised [us]. I asked her if she could see Liz too— she said yes . . . She ran to get a priest and a man. Liz saw, in a showcase, the poster about her memorial [service]—said, "Oh dear, I didn't know—that explains so much." We were in a room, I try to hide her from a nun and a priest, there were curtains around the area. Liz said, "I need to go back." I cried, I said, "No—I need you and miss you." She began to fade

away. *The curtains began to close. I cried, "I want to go too."
She said, "It makes sense now ... No, I must go back alone.
You can't come yet." The curtains closed and fell and split.*

GARY MAHMOUSSIAN, 24, BODY BAG #44

Uneven, gifted, lonely, and young, Gary is dead. Buried under
a bronze plaque in Heavenly Gates Cemetery to the west of
town past the tracks—near McDonald's—he can't answer his
own questions anymore so others must try for him: his sister,
her husband, his buddy, his widow. Tack down the basics:
childhood, adolescence, college, marriage, work. And hazard
guesses at the rest.

November with winter blowing in off the plains, cold
rain on joyless streets; the kids are in libraries, not on the
lawns—a college town with exams coming up; the workers
stay home at night and watch television; half-empty bars—
a factory town before Christmas: Terre Haute, Indiana.

Sharon Jaeger, Gary's sister, and Lewis, her husband, talk
to me in different coffee shops on different days; they are
separated now, and both of them say the crash has had
something to do with that. Their lives and Gary's were in-
extricably entwined, and so are brought together here.

There was a "real" father whom none of them truly knew,
Patrick Hills, a pilot in the navy's Hell Diver squadrons during the war. Dead at the age of thirty, in 1953.

Lewis: "He got killed *by* the war. When he came back
he started having a lot of emotional problems, physical problems 'cause of the strain. A lot of dive bombing tends to do
that to you ... I guess it was one of the worst things to be in."

Sharon: "I think it caused some of his mental problems.
He would very rarely talk about his experiences, according

to my mother. For a long time, she told me [he died of] a heart attack, because that's an easy way to explain it, but it was actually a blood clot in his legs and it moved up to his heart. He was the kind of person who would never go to a doctor. And I guess he started gambling before."

Family secrets.

Sharon: "Our stepfather married my mother when I was six and Gary was four. I don't know if it boiled down to jealousy, but my stepfather was really hard on my brother and he had no reason to be. A lot of it stemmed from the fact that he was from the old country where children respect their parents, especially their father—no matter what."

Aram Mahmoussian emigrated as a young boy, in the 1920s. He married late, at forty-two, and found himself instantly a father of two.

Sharon: "If my stepfather had been cooperative, there wouldn't have been any problems. Now, he was fine until Gary and I got old enough to have our own opinions and our own minds. As long as we were little kids that did what he wanted us to do, fine, no problems."

There was a year's stint in Miami when Mahmoussian opened a hamburger stand "on a busy highway with people who didn't want to bother taking the time to get off. It didn't make it and I think he declared bankruptcy, I'm not really sure. Gary and I had money from our real father— we'd been getting money from the government every year, and it amounted to quite a bit of money, which was supposed to be given to us when we were twenty-one. Well, my mother let him have this money to start this business, so there was nothing left."

They all moved back to Terre Haute, but the tension did not ease. "Gary made fantastic grades, he was very talented, he never got into any kind of legal trouble, but he finally got to the point where he just wouldn't take it any more. So our lives consisted of locking ourselves in his room and turning on the stereo."

Within a year Gary started his own band, and it was the

"best thing that ever happened to him." The band, first called the Snake Charmers and then The Chain Gang, became the nucleus of Gary Mahmoussian's life. They played at fraternity parties ("A big thing, you know, for high school kids"), at sock hops ("I was the chairman of our class dance committee so of course I always had them"), at concerts, in friends' living rooms, everywhere. The Chain Gang became somewhat famous locally and attracted a following, including Sandra Winchester, the tall, clear-eyed daughter of a dean of the university.

Sharon: "Sandra was in the class between Gary and me. I was a real good friend of hers for a long time. It surprised me when they finally got together. I just didn't think that they were each other's type, I guess. I don't know."

Lewis: "Sandra and her girl friends used to hang around the band all the time. She was going with somebody else, but Gary was always around—they were sort of like always around each other for ten years."

The short, skinny, blond kid with the funny Armenian name, the townie, the "stony" (as they call them here because of the limestone quarries); and the cool, beautiful academic's daughter.

Staying away from home as much as he could, working at ITT and then at ABC Electronics, Gary wrote his music and went to school.

Lewis: "He was really particular about his music instead of just silly . . . you know, just writing a song and remembering it to play. He had sheet music written for all of his music and had recordings done of it—he had about four hundred songs altogether—studio recordings actually, in big studios.

"He was very sensitive. When you hear the music, you can hear those things . . . But he was sad a lot of times . . . melancholy. I never saw him look jovial. He got beat down a lot by his father. He'd be successful in school—he had a 4.0 average his first semester at the university and his dad told him if he had two semesters of a 4.0 he'd give him a car. He wouldn't let him drive the family car; my dad gave me a

car, you know, and I never made straight A's. So Gary had
some problems at home and got this chance to go to Florida
for two months and dropped out of school for a semester. He
came back and made another 4.0 and his father wouldn't
give him a car because they weren't consecutive semesters.
But he didn't tell him this before, and still he wouldn't let
him drive the family car. Here's Gary, nineteen years old
and without wheels at all."

Important things in a young man's life—wheels.

Gary whizzed through college, worked nights at the electronics plant, finally bought a car—joyously succumbed to
writing a song about it.

He and Sandra circled, broke off, made up, graduated
Phi Beta Kappa a year apart. Sent to Washington by ABC
Electronics, Gary returned to Terre Haute to get married.

Sharon: "Knowing them, I would imagine it would have
been a spur of the moment thing, because if they had both sat
down and thought about it for a long time, they probably
would have scared each other off."

Together they moved East. Sandra didn't last long in
Washington; a convenient offer of a job back in Terre Haute,
at the university, brought her home in early fall of 1974.
Gary was due to follow that winter.

Lewis: "I loved him, admired him, but he would have
been a horrible person to be married to, because he liked to
be alone and he liked to work more than he liked to socialize.
He was an achiever, but he hadn't figured out yet how to
make it work.

"All you have to do is find some time in your life when
you want to work your ass off for about four years and you
do it. And get that out of the way and you get there. There's
no way you can avoid it. I don't care if you're not competent
at all, once you get your shit together and work at it for four
years, doing something, it doesn't take that much of your life
to dedicate yourself, but you . . . you just got to sit down
there and do it. It's your little prison term in life. Like Gary—

he wasn't particularly interested in staying permanently in electronics, but he knew that to make anything, you got to have a base, and you got to do it.

"We talked a lot about it. And it was just that he wanted this job in Washington and he rejected the fact that it took all of his time. It was advancement. The pay was really good and he got promoted. The reason he was coming back here was a promotion."

Sharon: "I'm sure he liked his job to a certain degree, because Gary would never do anything that he abhorred. But it seemed that he knew that he would eventually get a higher position, a desk job, which is what he wanted. He didn't like the manual labor [in the electronics factory] that he had to do.

"He worked at a blue-collar job but he wasn't a blue-collar person . . . I think he wanted to get to the point where he was self-sufficient enough, making enough money so that he could really get into music. That was his life. I don't think he would have ever given it up."

And Sandra.

Hearsay, guesses. Lewis: "I don't . . . Now this . . . I have the feeling that something was happening with Gary and Sandra because I . . . If I was married to somebody, I couldn't, no matter how good the job was, couldn't live away from them for four months knowing that you're not going to be with them from September to February or something like that. I'm pretty sure she . . . They all claim it was like old times, but you can only go out dancing so many times with different people without getting close to them—just being around them and not being around who you're married to. So I don't know, I had a feeling it was about ready to . . . Of course, I think your feelings tend to . . . when somebody dies . . . you reinforce your love for them 'cause it's easier."

Sharon: "They were, or at least appeared to be, so much in love, they were always right next to each other, ate off the same plate, always had their arms around each other. I was threatening to get them a scarf that was long enough to go around both their necks because they were always right next to each other."

Gary came home for Thanksgiving. He did not see his sister ("We didn't need to see each other, if that makes any sense; we were close enough to know how each other felt, we didn't need really even to communicate"), but he did see Lewis. They talked about an electronic contraption they were inventing—a "computer inventory control system," which "some people from Honeywell had come down to talk with us" about—talked about life and, says Lewis, death. "He was really comfortable with the idea of death; he made the comment that, 'Well, one thing about dying is that you don't know you're dead.'"

But by dusk on Sunday December 1, 1974, the others knew.

Sharon: "It started out to be a beautiful Sunday. It was snowing and I love the snow and everybody was really happy, and we were going to take the kids to a movie, and we were just getting our coats on and Lewis came up and said something about there'd been an airplane crash. And I tried to call Sandra and nobody answered. And I called her parents and nobody answered, and I thought well, that really doesn't mean anything."

Sandra had driven Gary to the airport in Indianapolis, an hour and a half away and "her parents were just out. I have a feeling we heard it all on the radio at the same time. Sandra was in the car driving her friend somewhere and my parents were on their way down to my real father's parents in Florida and it was around noon. And I just kind of put it out of my mind. Those things just don't happen. So we went ahead and went to the movie because I didn't want to sit around home worrying about it. What was it? Oh, *Robinson Crusoe*. Really high-class... But when we came out of the movies, one of the girls from the store was waiting out there with a note for us to call Mr. Winchester. And I knew right then. It was the longest walk of my life, from the theater up to the store. But when I called him, when he answered the phone he sounded okay. And so my first thought was, 'Oh, it's nothing, maybe he's just calling to let me know Gary wasn't on the plane.' But then he got serious, and so I knew. He said, 'We don't have the actual report yet, we don't know how many casualties there've been, but the plane Gary was on crashed.'"

What did you do?

"Cried."

Lewis: "We thought he was super when he was alive. I mean we always . . . I mean we all . . . it wasn't one of those things . . . He had a lot of people that thought more of him in life than they do in death. I mean he . . . he didn't get better because he died. He was somebody that you knew and that you built up while he was alive, promoted him because he was such a neat guy. And I don't know of anybody, that if I had to choose somebody whom I knew that would have to go, I'd say other than my wife and kids, I don't know of anybody that deserved it less than him."

Sharon: "I think Gary was always disappointed in me because I was satisfied with being a housewife and a mother, because I wasn't trying to make anything of myself . . . I just kind of felt it. I could be wrong . . . I thought so much of him that of course I wanted him to think highly of me."

Lewis and Sharon Jaeger separated eight months after the crash.

Lewis: "Right after Gary died, Sharon had just had her third baby . . . And she got really down. It was more than . . . it wasn't just Gary—it was just about things in general, and she got ten times more depressed than Sandra did. And it lasted for two or three months where she was just in a depression all the time. Then, all of a sudden she stopped communicating with me, stopped communicating with everybody, and she wanted to have a lot of real surface relationships with people. I mean I understand, I mean, I still love her, it's terrible ... and I miss not being around and there are things ... But she's just really cold now. I think she's afraid to have strong relationships with people—doesn't want to lose them. That has a lot to do with it."

Sharon: "Well, it was just my realization that here was somebody—Gary—who spent his whole life trying to attain a certain thing, and when he finally achieved it, his life was gone. I wasn't going to live the rest of my life for someone else. It was time I lived my own life. Do what I wanted to do. Which isn't easy, but I feel really good about it.

"I get really depressed a lot, but all in all I feel much better about myself. I feel like a real person now. I'd lived under my parents' rule until I was eighteen and I immediately got married and lived under somebody else's for nine years. You can't do that."

She has three children to care for, alone now. "I resent them very much, but I can accept it. I don't have much choice. Like I keep telling myself, it won't always be this way. When I'm thirty-eight, my oldest will be twenty, so ... thirty-eight's not that old ..."

Sandra.

Sharon: "Her philosophy as far as she reported it to me was that they'd had a good life, Gary lived a good life, she didn't feel sorry for him, she felt sorry for herself. Which is not the way I feel at all. She worried about me for a long time and still does, I think, because I'm still so emotionally involved in it. I don't know if she has accepted it and if she's really doing well, or just that she won't admit it to herself. I don't know which it is ... We went out on Gary's birthday and got stinking drunk, and we'll go out on the first [of December] again and tie one on. Maybe she's trying to tell herself that he wasn't really all that great. People rationalize tragedies in strange ways.

"Her father was in Europe when the airlines came around,

and from what I've been told he was really upset that she settled when he wasn't there, because I think she would have just agreed to anything they wanted. She just wanted to get it over with, sweep it under the rug."

Lewis: "Sandra isn't somebody who needs to be protected. Sandra's a very strong, a super strong person. If you don't know her she comes across as a little bit arrogant. But all of a sudden you realize that it's just because she doesn't need anything from anybody other than companionship."

Sharon: "Sandra is a strange person. She doesn't like to cope with people she's not very close to. She doesn't want to have to put up with the hassle of not knowing what you're going to say, how it's going to be taken . . . things that most people don't even think about, but to her a big hassle."

Finally, Sandra herself. Her reluctance to talk about Gary is not entirely her own, but rather appears to be partly the influence of her father, the dean.

We meet, after a series of telephone calls. She is cool and judgmental as she sits behind a cluttered desk in her office at the university. She refers to Gary as "my husband," as "he," seldom by name. She makes the point that he was "a small man, three or four inches shorter than I—which would make him about five-five or five-six." She does it more than once. "But he was a very successful young man—he was twenty-four and making a lot of money." But money does not seem to be the kind of commodity Sandra Mahmoussian is really concerned with. Perhaps she thinks that's what I want to hear. "I feel no need to . . . express frustration or anger" at his death, at the airlines, she says.

She has produced a record, culled from songs which Gary had written and recorded, and had seventy-five copies cut. "With the album I feel I've discharged my . . ." she searches for the perfect word, comes up with "duties"—"discharged my duties to his friends." We go, together, to listen to the songs, at her parents' house.

Her mother says she can't bear to hear the record and leaves the room. We sit at the dining room table looking out over a small lawn. The wind strips the last yellow leaves from an oak. Sandra hums with the music. Sunlight sporadically drenches the tree. She looks out, dry-eyed, at autumn's end.

The record is over. We drive back to her office. She gives me a copy of the record. "I'm lucky," she says. "There aren't many people who have five or six reels of their husband on tape."

BENJY APPLEWHITE, 3, BODY BAG #11
SUSAN APPLEWHITE, 28, BODY BAG #93
JAMES APPLEWHITE, 32, BODY BAGS #102 AND #102-A

> *First the sting; then the numb; then the*
> *unbearable pain of sorrow; then gradually,*
> *ever so slowly, the settling of a sad*
> *memory which from time to time will*
> *not allow you not to cry.*
>
> *Sue, Jake, and Benjy,*
> *My heart has been broken before, but*
> *never more than this.*
>
> —A MEMO FROM REPRESENTATIVE
> ANDY JACOBS TO HIMSELF

Sitting in a corner of Representative Andy Jacobs' front office in the Longworth House Office Building on Capitol Hill in Washington, D.C., Tom Hipple, the Indiana congressman's administrative assistant, is crying. He is crying openly and he is not ashamed. Three people who were his friends are dead. He is crying because he loved them.

On the floor of the House the members are debating a $116.1 million appropriation for further solar energy and environmental research.

"I couldn't really not like Benjy because he really liked me. He always called me 'Hipple'—I don't suppose that

Benjy knew my name was Tom, and that's because Jake and Suzy and a lot of my friends just called me 'Hipple' . . . Benjy would be playing and you'd be walking down the street and you'd look up and see him—he could get just about to your knees and when he'd hit you, you really had to be ready for him—and he'd just run up and throw his arms around you and then he'd walk carrying on a conversation with you back to the house.

"By the time Benjy was three years old—I don't remember how old he was when he died, I guess he was three—by the time he was three he was doing phenomenal things. The best description of it is that he was the translator for every little kid on the block. Most three-year-old kids can't communicate with an adult. They'd mumble something in little-kid talk and Benjy would say, 'He wants a cracker' or whatever. He could span both gaps. I hope I can raise my kids that way. I mean, I swore it wouldn't work. The kid was a terror when he was one, one and a half—Benjy was never told 'No.' He was always spoken to as an adult from the time he was two months old. It was never this 'I'm your mother and consequently do it,' that sort of thing—but Benjy is the one I really feel bad about because he had an opportunity to develop the way you really want the next generation to develop. I suspect every generation says, 'Well, I want to leave the next generation better off,' and there was a kid that really had the opportunity to do some things . . . My wife hears me say this every day. I . . . I feel bad about Jake and Suzy but it really hurts about Benjy."

After moseying through Indiana University, James "Jake" Applewhite went to Europe. "He worked in an army laundry in Germany; he was a bricklayer for a while. I'll tell you the truth, I don't really know too much. I've forgotten a lot of it, plus I'm sure there's a lot of it he never told me as how he got along because he knew I would worry. He said his mother always thought that you had to sleep in a modern house, in a clean bed, and get up and eat a balanced meal."

Lariana Harbison—Jake's mother, remarried since Jake's father died—still lives in Bloomington and still works for RCA, the big company in town.

"When he got back he got married. He married a doctor's daughter, a girl that he had gone through school with and they had what we thought was a mad love affair, but after three years they were divorced. They wanted different things. He sat down one day and said, 'Mother, there must be more to life than living in Bloomington, joining the country club, and running around with somebody else's wife'—which of course in a town this size there's a lot of.

"He went to New York City and the first thing he did was get a job with the Welfare Department. He was going to just do great things."

("Everybody in our age group," says Tom Hipple, "should have ended up in the East Village at one time or another.")

"He went to New York," continues Jake's mother, "and was going with a real fine young girl there who was a graduate student. Then he met Sue at one of the parties, and he started writing about what a cute little Chinese girl he had met. But, he said, 'Don't get uptight, Mother,' 'cause he knew Mother would. He said, 'we're just friends.' And the first thing I knew, it was very serious. And poor Harb—he's had to take me through so much since we've been married." Merrill Harbison smiles from behind a huge green cigar. He used to be a farmer.

"Since we've been married," says his wife, "I've lost my brother with a heart attack, my mother after a long illness and Jim—" she called her son Jim—"Jim's marriage to a Chinese girl was one traumatic experience. I met her when I went to the wedding . . .

"I've led a pretty narrow life compared to Jim. I have not traveled, I did not go to college. I've lived in Bloomington for thirty-three years. I came here just before I had him and I haven't lived in a, well—there are a lot of foreign elements on the campus, but I have not, I've never known an Oriental in my life. And, of course, with my husband being killed in World War Two, I just automatically disliked Orientals even though—"

He was in the Pacific?

"No, he was killed in this country in a plane crash and Jim was killed exactly thirty years [later] to the day, December first. His father died on the West Coast at night on December first and he died on the East Coast in the morning, which," she says very softly, "I think is quite a coincidence. Just unbelievable."

. . . and you went to New York?

"We met Sue and of course loved her. She was just a very lovely person and she and Jim were so compatible, just . . . She really was a big help in quieting him down . . . because he was, I'm sure, not the easiest person to live with. He had his ups and downs . . . Jim had so many dreams. He was not realistic in many ways and she went along with it, to a point. Yet she knew when to guide him. She was really good for him. And we had a marvelous time at the wedding . . . in the chapel of Columbia University. She was working and taking some courses there . . . It was the damndest service, a combination of a Chinese and a caucasian service and they read—of course Jim wrote poetry and Sue did, too—and they read some of their poems.

"Everything that could go wrong went wrong. In the first place it was such a mixed crowd. Of course Jim had many black friends and of course many, many Orientals because she's from a large family—her mother had been in several businesses, but by the time the kids were married they had just a kind of general store in Chinatown and lived in a small apartment where the children were raised. They live, like so many people that come from foreign . . . They didn't have any fancy furniture or anything like that, but they probably had more money than most of us dream of having. . . .

"Of course I didn't know any of them and then we left the chapel and went to a Chinese restaurant that belonged to an uncle of hers in downtown New York and a sit-down, thirteen-course Chinese dinner for two hundred people and I naturally thought that I would be sitting with my sisters and instead of that, I had to sit with the family, her family ... whom I'd never met until then, and of course, the older ones spoke very little English, and I tried eating with chopsticks. They finally got me a fork.

"It was such a shock to me ... I was so upset to think that he would marry that little yellow girl. Then I found out that her mother was even more upset, if that's possible. Naturally I would think anybody would be tickled to death to get my son, but of course she didn't like the looks of that—'big hairy white man' as she described him. She was very upset. But of course Jim won her over like he did everybody."

With Susan eight months pregnant, Jake moved to Washington and got a job (just in time to have Benjy's birth covered by the government health insurance plan) working in Andy Jacob's office.

"I had nothing for him," Rufus "Bud" Meyers remembers, "but he'd come into the office every day and sit around. He made us know him. Finally we said we have to get this guy a job and gave him a temporary one, thirty days at a time, as a legislative correspondent."

Meyers, now Representative Barbara Jordan's administrative assistant, then held the same job for Jacobs. He is a burly, bearded black man with a huge smile who says Jake "wrote really well. He was very sensitive to what kind of answer to constituents' letters would appease them. And when he cared about something, he really got into it. When he didn't care about something though, he did a really shitty job.

"On the banjo, he played bluegrass, folk; he liked Cat Stevens, Bob Dylan, Jimmy Hendrix, B. B. King ... didn't have much of a voice, but he'd pick up the music ... after hearing it about thirty times." Meyers laughed. Jake was a friend. "In the summertime, he'd sit out on his porch and play, and people would drop by and sit in the yard and listen —that was when he lived at Third and C Streets . . .

"He was too skinny for sports, except softball. Drinking beer was his only real exercise. Suzie didn't drink, but she was never down on Jake because he did. Sometimes she'd come with us to Whippy's with Benjy, and then she'd take him home to put him to bed and come back. We talked a lot of politics, a lot of Hill shop talk. Suzie always wanted to be in on the conversation and could get pissed off when she felt left out." Meyers shrugs. He knew her pretty well; she worked for him in Barbara Jordan's office, during the last months of her life. "Maybe it was because of the rigidity of her up-bringing—where she was 'just a girl' and was to remain silent —but she really didn't like being left out.

"One time Jake was having a rough time with Suzie, and I was just going through my divorce, and I said to him, 'You don't want to do it this way, being six hundred miles away from your kid.' And Jake said, 'Okay, I'll straighten myself out, you're right.' And he did."

Tom Hipple suspects that Jake and Suzie came to Washington in late 1969, because "he wanted to write and wanted to play the guitar and the banjo, yet he needed to do something a little more committed for a while."

Hipple, who grew up in Indiana, beginning his political career managing a friend's campaign for mayor of Blooming-ton, guesses at Jake's interest in politics. "In an area like Bloomington there were active social-rights movement things going on. The first three people, the first three young socialists, the Lions people, were rearrested, and the crackdown on the New Left was not at Berkeley, but in 'sixty-two in Blooming-ton. So there was political activity there. Jake was a very sensitive and intelligent guy and I think there was a real social type pressure that involved you. You know, the civil-rights movement was a social club and everything, and so it seems very easy for somebody to go into that sort of thing. Who's going to turn down a bus ride to Selma, Alabama? I mean, that's much better than going to the state tourney; I mean, you've seen a lot of basketball games, but that's something new. So I see how it can be done, and the environment was certainly there. You had the intellectual community, the university community—you could get a busload of people from Bloomington. It makes sense.

"Also," he laughs, "he needed to have Benjy on the insurance. I think they made it by something like seven days and nine dollars or nine days and seven dollars. He was married, he needed a job, Benjy was coming, and Andy [Jacobs, in whose Congressional district Jake was raised] certainly is a good fellow to work with and for. Philosophically and emotionally, Andy's crazy. I mean, when you compare him to the rest, when you determine that crazy means somebody out of the norm, and you take the norm of the four hundred thirty-five individuals that make up Congress, Andy's certainly out of the norm. But who'd want to be with sane people?"

When Jacobs lost his House seat in 1972, Jake was out of a job.

"During the month or so he was out of work," recalls Bud Meyers, "he wouldn't even apply for jobs known to be open with people he didn't respect—Senator Vance Hartke for example. 'Hell, no,' Jake would say, 'I'd rather starve.'"

Through the old-boy network of the House and the marine corps, Jacobs, who had served in Korea with Representative Pete McCloskey, got Jake onto the California Republican's staff.

Bob DeVoy, McCloskey's administrative assistant, says, "Jacobs was one of the few people whose requests McCloskey would always honor. But, in my book, people who walk in that way always have one strike against them. Jake got rid of that strike very quickly. He, first of all, wasn't ashamed of having gotten the job that way. He knew his way around the Hill well enough to know he was useful." Yet, says DeVoy, "Jake's areas of responsibility with McCloskey were not things he was inherently interested in—economics, business, wage and price controls . . . things like that. But he was a good writer, though he'd much rather draft a letter which would go out to five thousand people than answer one individually. Sometimes, to alleviate the tedium, or his own frustrations about it, he'd draft a really funny, really witty letter, sometimes it would be a little put-down to some constituent, and we'd have to get it rewritten. But though Jake could have a real Indiana folksy manner, he was totally comfortable with constituents dropping in or high-powered business executives from the Santa Clara Valley electronics firms, or people from the Stanford faculty."

DeVoy was both puzzled by and grateful to Jake and Susan Applewhite. "I came away so much better for having known them both." He explains that he was from a small Nebraska town, went through private schools, got his law degree from Georgetown, and was, by his own admission, stiff, and a little rigid in his upper middle-class values—the kind of person, he says, whom you would expect to find working in Pete McCloskey's office. "Jake was unique in here.

"I've never talked this way about him—or anyone. I cried at the funeral. I never cry. I went out to my car after the liturgy and wanted to go home or up to Jake's house . . . When he was alive, there was never any beginning, middle, or end to an afternoon at Jake's. If he was going out, he wouldn't hint for you to leave. He'd say, 'we're going to see so and so, why don't you come?'

"The Applewhites made you confront yourself, but not in a threatening way—just through the way they lived. Jake never compromised on friendship—you could share your depressions with him and your highs . . . I remember after one softball game—McCloskey's staff's slow-pitch softball team was one of the worst in the league—and it was the day a CBS

crew was filming 'One Day in the Life of Pete McCloskey'—
the two summer interns gave a party in a ramshackle Wis-
consin Avenue house. Sue left early and," says DeVoy with
wonderment, "Jake stayed. We finally walked back to his
house. It was late; he was stoned and he had a huge wide tie
with an American flag and it came down to his knees, and a
funny hat. And his pipe—he smoked a pipe all the goddamn
time and had all kinds of funny little instruments for it...
We walked back together ... on a hot summer night.

"Jake," he says, visibly moved at the memory, "never expected anything of you that you weren't." But McCloskey,
DeVoy says, had a habit of asking his aides whose views ran
contrary to a subject to deal with precisely those matters—
Catholics with abortions, Jews with the Arabs.

McCloskey accepted to sponsor Stanford professor William Shockley's request to hold a seminar on the Hill. DeVoy
says, "You know, the fellow who contends that non-whites
are genetically inferior to whites. McCloskey asked Jake to
write the 'Dear Colleague' letter informing the representatives of the seminar. Jake, who felt Shockley's contentions
shockingly attacked the intellectuality of his wife and, mostly,
of his son, had two or three, uh, *meetings* with McCloskey,
and finally agreed to write the letter on the condition that
he would be allowed to rustle up people with the opposite
point of view who could come to the seminar. He did, and got
four or five young, bright blacks associated with the NAACP
who came to the seminar and who rebutted every one of
Shockley's theories." DeVoy smiles, remembering the incident. "Jake quit shortly thereafter—something he'd been
planning to do anyway."

Stony Cook says he first met Jake two days after Representative Andrew Young of Georgia was elected, in November
1972. Young's administrative assistant, Cook came to Washington to settle in "and I wandered into Jacobs' office next
door and introduced myself. Jacobs had just been defeated

and his people were packing. Jake said he was looking for a job. I said I had nothing at the time—we had a lot of prior commitments to people in Atlanta—but asked him to give me his résumé and to keep in touch.

"I ran into him about a year later and offered him a position. He was the kind of person we needed. He had energy, he was a writer, he liked the members and they liked him. Most people around here just have jobs. They're not here because they dig doing them. Not Jake. He worked for less money for us for five months—we have a big staff—than he was offered elsewhere.

"He pulled a lot of Saturday duty, because of Andy's policy of having an open door on weekends . . . And he had a healthy relationship with Andy—they were really beginning to get to know each other . . . He lived with me down in Atlanta in the summer of 'seventy-four when Andy was running for a second term. You know, one of the problems of getting out the black vote was transportation. Jake devised a system, using a computer, so that on election day you could see the returns and determine who needed prodding—and provide wheels."

Hipple believes Jake took the job with Young "because they were involved in maintaining the superstructure of the civil-rights movement, when it was obvious that the movement had lost a lot of its public play. But there was still a need to maintain people in leadership positions because they're gonna be needed again. I think that the project that's fallen on Andy Young is to keep Martin Luther King's organization together enough so that someday they can come alive again—hopefully very soon. Jake was involved in that area and really had a fulfilled feeling because of it."

When Andy Jacobs was re-elected to Congress in 1974 he naturally offered Jake a job. Jake's reply, in a letter dated November 14:

214 SOUND OF IMPACT ─────────────────────────

Congressman:

Enclosed are a couple of items you might want to glance at in view of your interest in communications and commerce [two of Jacobs' House committees].

Aside from sending these on, I wanted to lay down a little personal note—to wit, how much Sue and I have always valued your friendship and the lasting gratitude we will always have for the crucial help you gave us at a rather difficult time. I'm sure you know this, but I don't want any chance of having anything like employment be misunderstood.

I sincerely appreciated your mention of that on Nov. 6th, and I had previously given much thought to it. Frankly, I expected it would happen, but I thought it might be more of a courtesy on your part—something which in no way diminishes my appreciation of it. The fact is, Andy, I hope to be working with you as much as possible, but not for money this time. You began with us as probably our most admired person in public life, and then as employer. That soon became friend, and that's a heavy word with me. I'm hoping it can become even more so by relinquishing the element of dependency— at least on anything but human friendships.

I have some thoughts (& probably some prejudices) about things that could be done by you here and in Indianapolis in the next term, and I'm sure we'll have an opportunity to talk about things like that soon.

So much for you & I. I guess the reason I'm not leaving Andy Young for anyone else is that, through my experience this last year, I've had a chance, if not the obligation, to explore some old and deep personal things—not just the race thing most of us grew up with, but some other, almost spiritual facets of myself and things in this country. The re-examination is often painful, but ultimately rewarding, I think. At any rate, it feels like something I gotta follow right now.

*Realizing this might all have been unnecessary explanation
. . . enough. I'll be talking to you before you receive this anyway.*

Onward,

Jake

Three or four days before going home for Thanksgiving, Jake
turned to Stony Cook and said, "I was born a white racist—
any time you find me or my actions steered by that, you let
me know." Cook allows himself only a quick blink as he remembers Jake's admonition before saying, "I got the feeling,
though, that he was reaching the end of his time on the Hill.
He'd bought some land in Canada and wanted to move up
there. First he talked about buying the land—" he smiles, no
doubt thinking of what must have been Jake's endless happy
talk about it—"then about borrowing the money to buy it,
then about getting a four-wheel drive vehicle, which is what
you need up there. When he was down in Atlanta, we even
went and looked at some four-wheel drive machines."

"Jake was a very talented guy," Tom Hipple says, "who
could have moved to Canada and kept his family and himself
together by being a farmer or, and I expect if Jake had had
his way, he would have played background guitar three
months a year and then practiced in Canada the rest of the
time . . . Jake was just about ready to find out who he was;
he was just on the verge of breaking loose, be it writing, be it
the guitar, be it whatever, he was just . . . and I feel the same
way. I'm thirty now and I've just really—all of a sudden,
things are falling into place. I think that Jake had the same
situation. Things that you've always wondered about, couldn't
quite understand, make sense now."

Jake's mother smiles indulgently, remembering the land in
Canada. "Old practical me . . . They didn't own a home,

they really needed nicer furniture, they really needed to trade cars—I could see all kinds of things they needed to do." Borrowing money, and sweet-talking a bank loan officer, the land was bought in Quebec Province. "And Sue wanted to work because they were paying two hundred forty-seven dollars a month on this land. Of course she had to go to work to help pay it—and there they are with a child ... This is all they wanted to do. And then of course, Jake was such an enthusiastic person. He showed me pictures of the land and he just ... you couldn't help getting enthused for him. Even though Harb—every time we talked about it Harb would look at me ... Harb was raised on a farm, he knows what hard work is. But when you hear the things they were going to do ... build a house themselves, you know. But anyway, Jake was so enthusiastic and I just accepted the fact that he was happy, he had a dream ... Well, I don't think it was a dream—I think they would have done just what they were planning. They would have paid for the land, they would have quit their jobs, they would have gone up there, and probably, after a year or so, they would have both realized that it wouldn't work and they—" Mrs. Harbison shakes her head. "Jim said, 'Mother, we can always work.' Well you see, there again—I went to work during the Depression and I have stayed on this one job—I'm a secretary to the personnel manager at RCA—thirty years. Jake said, 'Mother, you need to get out of your rut,' and I said, 'Honey, once I get in the comfort of a rut, I'm afraid to get out.'

"He told me one time that I didn't like a challenge. And I said, 'Jim, I'm all challenged out, I've had too many challenges.' But, as I say, I quit worrying because they were young, they were healthy, they were well educated, and I know they could have taken care of themselves."

Andy Jacobs walks by. He has frighteningly penetrating green eyes. He says he tried to get on Flight 514 but was told the only available seats were in first class. He waited for cheaper passage—after all, it was the taxpayers' money. On the wall of his front office, facing the hallway door, is a large framed black-and-white photograph of Jake, Susan and Benjy, up on their Canadian land. There is also a picture of John F. Kennedy.

"I was on the floor [of the House] on Monday [December 2]," the Congressman says, "and some man—and I must say this without prejudice—who was definitely a Dixiecrat said he'd read in the paper that Jake had worked for Young and that Suzie had worked for Jordan, and that he'd worked in our office. So this guy walks up and says, 'I read that there were some of your people in the crash. Were they black?' And I said, 'Well, man, one of them was white, one of them was yellow, and one of them was in between.'"

Tom Tipple and Jacobs nod at each other. Common indignation. And Jacobs, with whom the Applewhites had shared Thanksgiving dinner in Bloomington, now remembers hearing of the crash: "They had the names on television, on the screen. The thing that really hit hard about this was not just seeing the names, Mr. James A. Applewhite, Mrs. James A. Applewhite, but *Mister* Benjy Applewhite—that tiny little fellow. It was a dirty lie because it was not *Mister* Benjy Applewhite. He hadn't gotten that far in life. It implied that he was a guy on the plane, that this little fella had maybe made love, or been places, had known what life was—but he hadn't."

ON TRIAL FOR WRONGFUL DEATH

The lawyer beckons the insurance executive aside. "Give me two-hundred-and-fifty thousand for this case," he mutters, "and you'll get me out of court."

"Fifty-thousand is my offer." The insurance man looks down at the lawyer. "I knew you when you were still trying cases in small claims court in Brooklyn."

"I've grown since then . . ."

"Only your ego," says the insurance man.

—CONVERSATION OVERHEARD OUTSIDE A COURTROOM

"Members of the jury, you have today a trial for a civil suit for damages, and damages only, for the death of a person who was killed in the TWA airline crash of December 1, 1974. The crash occurred near Upperville, Virginia, as the plane was approaching Dulles Airport for landing.

"The name of the decedent—that is, the person who was killed in the crash—was Annie Mozelle Killingsworth. And she sues through her administrator, Mr. Jackson Rose. The estate, that is, the estate of Miss Killingsworth, is the plaintiff in the suit, and the defendant in the suit, that is, the person who is sued, is Trans World Airlines. Now, Trans

World Airlines has, for the purpose of this suit, accepted full responsibility for the decedent's death. Therefore your function will not be to determine fault in this case, but to determine what amount, if any, of compensation should be paid to the beneficiaries of Miss Killingsworth's estate."

Monday, March 1, 1976, is a prematurely warm spring day, and the growl and rumble of jets landing at National Airport and making their final approach down the Potomac five blocks away rolls in through the open windows of the U.S. District Court in Alexandria. Judge Albert V. Bryan, raising his voice to be heard, asks, "Are there any of you who have any belief that it is improper to sue or make a claim for money damages for the death of a person as a result of the negligence or improper act of another person or firm?" He peers at the prospective jurors sitting four-deep in the back of his courtroom. Two-score of pinstriped backs in shades of gray and blue twist to assess the reaction to his question. The panel of impressive Virginians remains silent, and the assembled attorneys—present in force to scout TWA's defense and their own colleagues' tactics—turn back toward the Judge.

"Members of the jury," says Judge Bryan, "the plaintiff decedent was black. Her beneficiaries are black. Would that affect any juror's ability to render an impartial verdict in the case?"

Blank stares. ("I would never know a juror who would answer in the affirmative to the question," the Judge told Aaron Broder, the plaintiff's attorney when, during a brief bench conference, he requested that the question be posed, "but it may do well to at least have them think about it.")

So far, the families of thirty-four of the eighty-five crash victims (relatives of the seven dead crew members are covered under the Federal Workmen's Compensation Act and cannot sue TWA) have already settled their claims against the airline, the majority without even threatening suit. Wrongful deaths suits have been filed on behalf of thirty-nine of the remaining victims and are beginning today, scheduled for trial in rapid succession. Neither TWA nor the multitude of plaintiffs' lawyers actually expect all the cases to get to court. How many will do so depends largely on the outcome of the first few trials. Should the jury awards be modest in comparison to Associated Aviation Underwriters' offers, there will be a rush to settle. Should the juries liberally interpret Virginia's new wrongful death statute (which puts no ceiling on potential recovery for mental anguish, sorrow, and loss of companionship and comfort) and return awards substantially greater than AAU's settlement offers, the plaintiff's bluff of waiting out a trial will have paid off.

By eleven o'clock, an hour after the proceedings began, a jury has been chosen—three men, three women; all white—and the packed second-floor courtroom is hushed as Aaron Broder, sweating slightly, rises to argue his case on behalf of Annie Killingsworth's family who, six-strong and dressed in mourning for their day in court, sit in somber row behind their counsel's table.

Broder, F. Lee Bailey's law partner, begins by introducing the dead girl's mother, two brothers, and three sisters to the jury. On cue they stand, eyes downcast, when he calls their names.

Glancing at his notes, Broder tells the jury that Miss Killingsworth was twenty-five years old, an Ohio employee of the federal government's Department of Housing and Urban Development, who earned $12,841 a year and had a "very"—pause—"very promising future with the government."

He then introduces his remaining witnesses—friends of the family from Denmark, South Carolina; Miss Killingsworth's HUD boss—and, taking a deep breath, looks directly at the jury box to announce, "This suit is a suit for ten million dollars."

The jury regards him unblinkingly.

Hurrying on, he stresses how special, how exceptional Annie Killingsworth was—how "she had risen up on her own efforts ... to considerable accomplishment"; how she would have earned more than a million dollars during her working life; and how, though geographically dispersed, Annie and her family were "very close-knit," and her death consequently left a fathomless vacuum in her loved ones' lives; indeed, that her mother's "whole life came crumbling down" after the crash, "that there has been for her since then grief and suffering," that even in church, in the choir, "she participates in tears and in anguish."

Twenty minutes after he first rose Broder concludes by saying, "I only ask that you be patient, wait until the end of this lawsuit at which time I will come back to you and ask you for justice on behalf of the five sisters and brothers and the mother of Annie Killingsworth."

Judge Bryan nods toward the TWA counsel. "Mr. Booker?"

"May it please the Court, ladies and gentlemen of the jury. As Judge Bryan has already pointed out, I am Lewis Booker. I am from Richmond, Virginia. Virginia Powell and Rand are with me. We are here today to represent Trans World Airlines."

Booker, a shortish, compact man in a perfectly pressed gray suit, speaks with a languid tidewater accent, unhurried, deliberately friendly. "What I am going to tell you now, of course, is the evidence which we believe will come forth from the witnesses on behalf of Trans World Airlines... What I say to you though, of course, is just what I think the evidence is going to be. It is up to you to decide what the evidence is and where the facts really lie.

"Mr. Broder has pointed out that this is a ten-million-dollar case. It is a ten-million-dollar case because that is the figure that the plaintiff chose to sue for. When one brings a suit in Federal Court, one pays his fifteen dollars and can announce any amount from ten thousand dollars to ten million or more. So that is not evidence."

Leaning his elbows and clasping his hands on the lectern set below the Bench, facing the jury box, he says, "I think I will start off by telling you what the evidence for TWA will *not* show. No witness on behalf of TWA is going to come before you and tell you that Annie Killingsworth was not the dutiful daughter and was not the well-educated person that Mr. Broder has outlined in his opening statement. There is simply going to be no dispute about that."

During a lengthy bench conference forty minutes earlier —out of the jury's hearing—Judge Bryan, to Broder's satisfaction, ruled that TWA could not admit into evidence Miss Killingsworth's common-law marriage with a Columbus man.

"But," TWA's attorney adds, "you are really not here today to determine what kind of person she was. You are here to determine, in dollars and cents, what TWA must pay to the administrator of Miss Killingsworth's estate because of this accident. And that is divided into two aspects. First is the question of grief and solace. I think we all know that there is no way one can suggest a dollar value on the death of an individual whom we love, and so I certainly intend to make no suggestions to you as to what that should be."

The assembled lawyers in the spectators' section glance at each other. Booker knows his job.

"The second aspect," continues the man from Richmond, "is one which is susceptible to monetary determination. And that is the expected financial loss which the claimants have sustained as a result of the death, and it is here where the testimony will differ greatly. We believe the evidence will show that none of the people who appear before you today were really dependent on Annie Killingsworth. They had jobs of their own. They had parents of their own to support them. They were perfectly capable of getting along without any income from her."

This is the heart of TWA's case. Booker's hint, veiled so far, is that in bringing suit, Essie English and her five grown children are merely trying to reap the benefits of an expensive law rather than recover damages for money and services of which they were deprived by Annie's death and which they neither needed nor expected; that they are, in effect, bit players in a scene written for and by aviation lawyers who represent cases such as theirs on a contingency basis, usually for a standard one-third cut of any settlement or jury award. How, had she not been sought out, could Essie English, a grade school teacher from Denmark, South Carolina (population 3400) have heard of F. Lee Bailey or Aaron Broder?

The plaintiffs, wedged together on a single bench, avert their eyes. Booker allows, tactfully, that the jury may find that there was "some actual financial loss," but, he says, that amount "will certainly not be a question of adding up all a person could expect to earn over an entire lifetime.

"So, then, what should be your verdict? Your verdict obviously is going to be to award some damages against Trans World Airlines. There is," he says meaningfully, "no way I can walk out of this case a winner. And you know that as well as I do."

Appreciative nods all around the courtroom.

"The question is, what is fair and reasonable?" Booker straightens from the lectern. "Mr. Broder has asked for justice for the family. I ask for justice not only for the family but for TWA. I ask for justice for both parties in this lawsuit, not just one of us." Silently nodding his thanks to the jury, Lewis Booker sits at the defendant's table.

Broder then calls the first of his eleven witnesses. She is Calley Wright, a friend of Mrs. English and her family, the wife of the church deacon and undertaker who buried Annie.

"Will you tell us, please, what kind of student she was and what kind of person that she became."

Mrs. Wright takes a deep breath and launches into her carefully orchestrated eulogy. "I would say that Annie Mozelle was brilliant, very perceptive and ambitious. She had a beautiful personality. She got along with all the teachers and her classmates, even through high school. She was very highly respected among her peers . . ."

Broder then has her describe how, following the crash, Essie English no longer was the joyous choir singer that she once had been and how, at Annie's funeral, "she collapsed almost."

Mrs. Wright is succeeded on the stand by her husband, Lucius, who corroborates his wife's testimony as to Mrs. English's breakdown during the funeral services (for which he had charged $760), and her apparent loss of zeal in life.

Booker begins his cross-examination by asking, "Mr. Wright, when was the last time you saw Annie Killingsworth?"

"When is the last time I saw her?"

"Yes, sir—alive . . ."

"Ann worked away a good while," says Wright, a dapper tight-mouthed man in plaid pants and a green blazer. "It's been a good while since I have seen her."

"Can you tell us, though, when was the last time you saw her?"

"No."

"Do you know where she lived?"

"Yes—Denmark, South Carolina."

"No, I mean while she was away . . ."

"Oh, not until the time of her death. I didn't know of her . . . while in school . . ." No eulogies here.

"Are you a funeral director?" Booker asks.

"Yes."

"Obviously you have attended a number of funerals? Is that correct?"

"Yes," says Lucius.

"On the date of the funeral for Miss Annie Killingsworth, you have described that Mrs. English was overcome by grief. Is that," Booker asks, "an unusual thing for you to see as a funeral director?"

Broder springs to his feet. "I object to that, Your Honor. It is not relevant or material as to other funerals."

"Objection sustained," shrugs the Judge.

"No further questions," says Booker, his point made.

Shortly before noon, Sonia English—the first member of the family to be called—is sworn in. Unsteady on her platform shoes, she walks by the jury without looking at them.

In response to Broder's questions she says she is nineteen, single, and is studying psychology and sociology at Fisk University in Nashville.

"Now," says Broder, "would you give us, please, the names of your living brothers and sisters?"

"Edgar, Brenda, Harold, and Dorcas."

"What are their ages, please?"

"Edgar is thirty-one; Brenda is twenty-nine; Harold is twenty-one; Dorcas is sixteen."

"If Ann had not died, how old would she be now?"

"She would have been twenty-six." Sonia swallows.

"Now, then—Ann was your closest sister in terms of age?"

"In terms of age, in terms of everything," says Sonia, begining to cry. "Ann was closer to me than anybody else in the world."

Broder waits for the tears to stop and leads her through a recitation of how Ann had helped her adjust to going away to school, to dealing with young men, to becoming independent. "I could depend on Ann to give me some type of courage so that it would help me adjust more to Fisk . . . I could identify with Ann, because of the kind of life that Ann had lived—an independent life—which was one I wanted to live someday . . . and I could depend on her that way. I couldn't depend on anyone else . . . She never forgot my birthday."

"How do you feel about Ann's death now?"

"It's like somebody took something away from me that I can't get back. Somebody took a part of me that I can't get back. I was—I identified with Ann. I was . . . *dependent* on her."

"You have cried on this witness stand," Broder notes. "Have you cried since Ann died?"

"Yes."

"Hard?"

"Yes."

"About Ann?"

"Yes."

Sonia is breaking down. Jurors and spectators shift uncomfortably.

"Do you have," Broder plows on, "any particular recollections about anything that happened to you on your eighteenth birthday, and its significance now?"

"When I was eighteen, Ann sent me . . . I feel it was a special card." Sonia fights back tears. "It had a special message."

"What was the message?"

"It said—the card said, 'God grant me the serenity . . .'" she chokes . . . "'to accept . . . the things I cannot . . . change.'" Sobbing loudly, she cannot go on. Broder, by the lectern, waits.

A piercing howl of pain bursts from Sonia's lips. Her foot comes down hard on the wood floor; it sounds like a rifle shot. Prisoner of some terrible, primitive grief, she rocks in the witness chair. Her amplified wails echo over the public address system.

Finally, a U.S. marshal turns off the microphone. The jurors look at their laps. Essie English rises heavily and escorts her daughter from the courtroom.

Broder watches them go.

"Do you have any more direct examination?" Judge Bryan asks.

"I do, Your Honor," says Broder.

The Judge icily suggests that he postpone it and call another witness.

Broder calls Edgar Tobin to the stand, Annie Killingsworth's older half-brother, and leads him through a smooth account of Annie's influence on his life: "She was more of a big sister. If you had any kind of problem, you could call her and she would sit down and talk with you and she would understand, you know. Because she just had the courage and the stamina to keep pushing forward . . . She used to tell me that I needed to smile, to change my personality to a more positive personality, that it would all come. These things began to help. Suddenly, it was tragedy [that] took her away."

"How often do you think about Ann?"

"I think about Ann all the time." Shoulders hunched under his blue suit jacket, his handsome narrow face in his hands, Edgar begins to cry.

The judge calls the opposing attorneys to the Bench. Turning to Broder, standing beneath him, and speaking distinctly but softly so as not to be heard by the rest of the courtroom, Judge Bryan says, "Mr. Broder, I know this is an emotional thing for the members of the family. Certainly on the issue of solace, emotion should be shown, but I am not going to allow you to bring each of these witnesses to hysterics."

"Now," coolly continues Bryan, "I am sure it is emotional, but the questions are apparently designed to provoke that. I don't want to minimize these brothers' and sisters' and mother's affection for the sister, or minimize the sorrow that they are undergoing as a result of her loss. But you accomplish just as much, it seems to me, without the detailed questions of how many times have you cried, how many times do you think of her."

Booker, listening in, keeps his face blank.

"If you don't stop it voluntarily," Bryan warns Broder, "I am going to stop it for you because I am not going to go through with any more brothers and sisters what we just went through with the young Sonia. Because they are going to have to get themselves together and testify. And you are going to have to help, because if you don't, I'm going to cut you off."

Chastised, Broder nods, respectful of the Court. "Your Honor, may I state that relative to the question of how many times have you cried, I think that goes to the question of damages in the case where the law has just explained its horizons to such an extraordinary degree . . . It isn't our fault. It's the legislators', in the sense that if they are at fault, there be—"

"I'm not talking about the philosophy of it," interrupts the judge. "I am just telling you if you keep after these witnesses long enough, you will bring them to the same point you brought Sonia to."

Broder is silent.

"Now," adds Bryan, "whatever that may be as a trial tactic, I'm not going to permit it. So you restrain your questions at a point. This last witness was just as effective with two or three questions as you will be by having him say ten times how many cards he got, and when does he think of her singing, and when does he cry." Bryan leans forward in his high-backed black leather chair. "You are overdoing it. Both with me, and maybe with the jury. So . . ."

Broder forces his voice to stay calm. "Unfortunately, the inherent nature of the lawsuit itself is tears."

"Tears I don't mind," snaps Bryan, "or I mind because I am distressed to see people distraught. But I am not going to have every witness brought to hysterics."

An uneasy truce is achieved. For the first time since the attorneys approached the Bench, Booker speaks up. "If Your Honor please, while we are up here—the Court has already ruled on the relationship of Sonny Thomkins. I understand the Court's ruling. However, as going to the question of the character of Annie Killingsworth, I would simply like to know whether I am going to be precluded from doing that in light of the Court's earlier ruling?"

It is a stab in the dark, but after witnessing Broder's up-braiding, Booker suspects the Judge might soften on the question of allowing testimony about Miss Killingsworth's common-law husband into evidence.

"I think you are precluded from doing that not only because of the Court's earlier ruling," says Bryan, "but because I think the Virginia Court of Appeals on the question of solace has ruled that sort of thing out."

"If Your Honor please," Booker says in his courtly Southern voice, "it is our view that once they introduce evidence of her estimable character, we are entitled to introduce evidence to the contrary. And, if they say there was no common-law relationship, then there is a meretricious relationship that I would like to explore."

Broder, on safer ground now, makes a stand. "I submit, Your Honor, that what he is attempting to do is not introduce a meretricious relationship but introduce a prejudicial element into the case. Whether this girl was living with a man at the time of her death in no way whatsoever would be related to any of the issues in this case."

The Judge does not interrupt him. Passionately Broder concludes, "I submit that any attempt to introduce that carries with it such prejudice and such an appeal to inflame the jury with regard to any testimony of that sort that it has no business in this case whatsoever."

Bryan sustains Broder's objection.

The Court recesses for lunch. In clusters, the attorneys whose cases are scheduled next fan out in search of restaurants. Broder's and TWA's tactics are reviewed. Booker's defense— evidently a model for the remaining trials—is impressive and Associated Aviation Underwriters' Fred Schaffhausen is, conveniently, on hand. He will close some cases this afternoon.

The insurance company's policy is to settle as quickly as possible, without trial, after an accident. "We want to get the whole thing behind us," one insurance man explained. "Early in the game we try to make generous offers. But once they take us to court, we get a feeling—a trial is a barometer—of how things go, and, often, we can reduce our offers because the jury awards themselves are less than we were willing to pay."

Insurance men tend to have an instinctive animosity toward big-name aviation attorneys who, they feel, often mislead their clients into believing they stand to collect huge settlements, thus easing the burden of giving their lawyer his 33-per-cent cut.

"In many cases," continued the insurance man, "people have completely wild and inflated ideas about the value of their cases. They read newspaper stories about jury verdicts—

which are probably on injury cases anyhow—where a guy becomes a paraplegic for life and has huge medical bills, or where a man dies and leaves a large family. Everybody nowadays tends to think in terms of millions of dollars and, at least at this point, that's not the way the business runs. If an attorney tells people what their case is worth, they may believe him because they hired him. If we tell them, they think we're trying to rip them off."

Edgar Tobin and Sonia English sum up their testimony during the afternoon session and, under cross-examination, both admit that their dead sister did not, in fact, contribute measurably to their financial support. They are followed on the stand by Broder's economist who testifies that, according to his calculations, Annie Killingsworth would have probably earned $1,009,000 dollars during the course of her working life.

Cross-examining, Booker asks, "Dr. Guilfoil, your calculations take no account of the actual amount of money she may have contributed from time to time to the support of her family?"

"That is correct," says Guilfoil, an associate professor of economics in New York.

"And this makes no—this . . . *calculation* makes no assumption whatever as to what amount of her income she had spent on herself? Is that correct?"

"That is correct."

"So this is just a gross figure. It has no deductions for her own expenses? Is that correct?"

"That is correct."

"Or for taxes?"

"That is correct."

"Or for anything else?"

"That is," Guilfoil says once again, "correct."

"And does this figure assume a hundred-per-cent probability that she would live to be sixty-five years old?"

"It does."

"And both those are variables," Booker says, "are they not?"

Guilfoil tries to slow the stampede. "I don't understand your question."

"Not everyone," says Booker, "at the time of twenty-five is going to be alive at sixty-five, is he or she?"

"No. Obviously she didn't live to be sixty-five."

Someone in the back of the courtroom coughs.

Broder's next witness is Paul Lydens, director of the Columbus office of the Department of Housing and Urban Development. He testifies lengthily as to Miss Killingsworth's professional qualities; the jury seems listless.

Mid-afternoon. It is Essie English's turn to take the stand. She is large, dignified, dressed in black. In her left hand she squeezes a white handkerchief.

After ascertaining for the record that she was Annie Killingsworth's mother, Broder asks, "Will you tell us, please, about your daughter before this accident occurred? Will you tell us about her from the time she was born with regard to her health, her habits, customs and manners, attitude toward her family, sisters and brothers, and toward you?"

Essie English nods. "She was born a very healthy baby. I never had any medical problems with her. She was always healthy. She was always cheerful, even from a baby. She was smart. She was quick to grasp things. She paid attention to most anything around her, even from being a baby. She learned to look at pictures in books very early. She was smart.

"She learned to walk earlier than most children do. She entered into the kindergarten early. She went to the Head Start. Then she went to elementary school. She was progressive all the way. She went through high school without any trouble. She was progressive.

"She made high scholastic records. She participated in extra-curricular activities and she was just a child that everyone catered to . . . She went away to college—we had a college in our town, but she wanted to go away to college. I told her if she was progressive in high school, I would do all I could to send her to the college of her choice. So she chose Wilberforce . . . She did well at Wilberforce . . . She was a Delta, which she took much pride in."

"What is that?" asks Broder.

"That is a Greek sorority. I am not in a Greek sorority so I don't know all of the meaning of it, because," says Mrs. English, "it has its secrets."

This is rehearsed testimony, delivered as though memorized, in a monotone, eyes nearly closed. But there is, just occasionally, a hint of a mother truly grieving for a dead child —not a perfect mother nor a perfect child, but real ones. And it trickles through—despite the bewildering coaching of a New York lawyer—in the gentle modulations of a voice belonging to a lady who has, all her life, sung in Southern choirs.

Broder asks her to tell the jury the circumstances in which she learned of Annie's death.

"I had been to church that morning. When I came home my daughter came into the house. She told me to take a tranquilizer and to sit down because she had something she wanted to tell me. 'Baby,' I said, 'Jesus is my tranquilizer. You just tell me what you have to tell me . . .'" She is crying. Alone in the witness chair.

"Sometime later that night, I guess between ten and twelve o'clock, TWA called me to inform me that my daughter . . . was . . . on that flight, and didn't survive."

"What did you do next," asks Broder.

"I passed out."

The Judge looks at Broder and, for a moment, the questioning becomes less personal. Shortly, however, Broder asks her how she learned that Annie had been found.

"It was Wednesday night. TWA called and told me that her body had been positively identified. I asked them, if the body had been positively identified, would this mean that she was in one piece?" There is visible swallowing among the jurors. "He told me he couldn't tell me. But he told me he would inform me as to when they would be sending her home."

"Will you tell us, please, did they send her home, or did you have to travel?"

"No. They sent her as far as Columbia, South Carolina, which is fifty miles away from Denmark."

"Fifty miles from Denmark?"

"Yes."

"That's where they sent the body?"

The TWA attorneys scribble notes to each other on yellow paper.

"That's where they sent the body," repeats Mrs. English, "and the funeral director picked the body up from there and brought it to Denmark."

"Did you open the casket?"

"No," says Essie English, "I wanted to see her but the family insisted that I try to remember her as a *whole* individual . . . I wanted to see if it was just an arm or a leg of my child. It would have helped. But I saw nothing of the remains of my child. Nothing. Nothing."

Broder glances at the jury, asks about the funeral and then, "When you wake up in the morning these days . . . how do you feel?"

"I wake up every morning crying for my child," Essie English says softly. "If I could just hear her say something . . ."

"Has time healed the wounds for you?"

"No."

"When Ann wrote to you, how did she sign her name?"

"When she wrote to me she . . . she would sign her letters, 'Love you lots, Ann.' She would call me all the way from Ohio just to say, 'Mother, I was thinking about you, and I called you to tell you I love you.' And now . . . now Ann doesn't call any more."

A long silence ensues. Broder slowly leaves the lectern and

walks back to the plaintiff's counsel table. Picking up a handful of folders, he tells Judge Bryan he wants to submit their content into evidence. The Judge calls a bench conference.

"We have," Broder tells the judge, "the Bible School—"

"Why," Bryan asks, "do you want to go any further with this witness? She has been an extremely dignified—heartbreakingly so—witness. It seems to me to go further is . . ."

Broder explains his point; Booker presents his side. The issue of exhibits is decided. Mrs. English, still on the stand, sits in silence between the jury to her left and the Judge to her right.

"Judge," says Broder at the Bench, "I am getting weary. I wonder if we could adjourn a little early after this witness."

Bryan asks him how many more witnesses he has.

"I have the balance of the plaintiff, but I am feeling awfully tired and weary."

The Judge shakes his head. "I'm sorry. This is much too early for us to quit. I want to finish the plaintiff's case today."

Beaten, Broder acquiesces.

Booker cross-examines Mrs. English.

"You are presently employed as a schoolteacher in South Carolina?"

"Yes."

"It is your intention to continue to teach school?"

"Yes."

"Your husband, is he employed?"

"Yes."

"And it is his intention to continue employment?"

Essie English glances at Broder. "I can't answer that."

"You testified," Booker smoothly continues, "that you helped out your daughter when she was going to college. Did you have to go into your own savings to help her out?"

"I didn't have any savings. I had to borrow money to send her to college."

"Then I believe you said that she was sending you money to—help you out of the hole, I believe was your expression?"

"Right."

"She was helping pay back the money which you had had to borrow for her?"

"Right."

"And had she paid it all back before her death?"

"No."

"How much was still owed?"

"Well, I didn't bring any figures or anything," Mrs. English says, "because I was not asked that question."

"Could you give us some general idea of how much you had to borrow in the first place, for her education?"

"I probably could get that from the bank," Essie English allows.

"But you don't have any idea now?"

"No. I don't have that with me."

"Was it in the neighborhood of one thousand dollars?"

"Yes."

"You borrowed about a thousand dollars?"

"It was more than a thousand dollars," Essie English says, the tears now long gone. She didn't expect this line of questioning and Broder can offer no help. "I had to borrow it to send her to college."

"But you can't tell me how much that was?" Booker, though unfailingly courteous, is dogged.

"No, not to the penny and I don't want to give you a wrongful figure."

"I would be happy to have any estimate that you feel comfortable with."

"No," insists Mrs. English. "I don't want to give you a wrongful figure. I would rather get the facts and present the facts."

"Can you tell us," Booker asks, "how much she had paid you back?"

"Not up to the penny," comes the resolute answer.

"Can you tell us approximately?"

"I didn't keep a record of it."

"How often did she send money?"

"Well . . . I would say from once to twice monthly, according to how her pay ran."

"How much was each one of these checks she would send you?"

"They didn't come in the same amounts. The amounts that she sent went all the way, I would say, from five dollars to a hundred dollars, but she didn't send all of that in checks. She didn't send all of it in money orders. She didn't send all of it in cash. She sent it different ways." Essie English stares at Booker.

"And the purpose of this was to pay you back for your being willing to borrow?"

"Not really to pay me, because I didn't charge her for what I did for her. She knew when she got out of college I was in a strain, because I was sending other children to school. She knew that. She wasn't trying to pay me back for what I had done for her. She was only trying to help."

"Can you give us any idea, during the year 1974, how much that amounted to?"

Mrs. English shakes her head. "I really can't give you any figures, because I didn't keep up with it."

By now Booker is plainly irritated. "Could you give us an *estimate*? Was it as much as five hundred dollars?"

Broder rises. "I will object, Your Honor," he says to the Judge. "It's already been gone over several times. She's stated she's . . ."

Judge Bryan agrees. He turns to Booker. "I don't think you can get any better estimate than you have gotten. Objection sustained."

Confident that Mrs. English's indifferent memory was not lost on the jury, Booker says, "Very well. No further questions."

Broder then has Annie Killingsworth's younger brother, Harold, and her sister, Dorcas, take the stand. Tearily they testify to Ann's invaluable help, guidance, and support. Under cross-examination, Harold—Booker declined to question Dorcas—is unable to recall just how much money Ann had given him over the years.

Down below, on South Washington Street, the rush hour traffic builds. Broder calls his last witness, Brenda Rodriguez, Ann's eldest sister. A handsome woman of twenty-nine, she is the only one in her family not to have dressed entirely in mourning. Instead, she wears a well-cut, black-and-white striped dress, and speaks in a clipped, educated, surprisingly un-Southern voice.

Broder asks her to tell the jury a little of her own background.

"I grew up in Denmark. I graduated from Voorhees High School in 1964. I went on to two years at Voorhees College. I left Voorhees in 1966. I attended Wilberforce University in the years '66 and '67 during which I got married. And after the two children—my youngest was three and a half years old—I went back to school and obtained a bachelor's degree."

Then, the by-now standard question: "Would you be good enough to tell us what your recollections are, what your affiliations and associations were with your sister Ann?"

"Well, I am the second of six children. The oldest is a brother, Tobin. Then myself. And the third child was Ann. And for a long time, we were the only two girls in the family. So I loved her very much. We had more of a . . . I guess a friendly relationship rather than me depending on her, or her depending on me. It was just like, maybe, a part of my body. You know, we grew up together. We shared things. We shared the same bed, the same clothes—just everything I can think of . . ."

Broder brings her up to December 1.

"So I went to my mother's," Brenda says. "When I opened the door Mother was—I had never seen her that happy in my life. She was so extremely happy when she opened the front door for me. And I haven't seen her that happy since. And I don't think she will ever be that happy again. I wish I could see her that happy again. You know, she was radiant . . . So I walked past her—it's an L-shaped hall— and I went into the kitchen where Dorcas was. I asked her to take my two children into her bedroom for a while, because I had to talk to Mother about something. So Mother followed me into the kitchen, you know, talking and laughing and whatnot. I told her, 'Mother, take a tranquilizer. I want to tell you something.' She wouldn't. I said, 'Well, I wish you would. I am not going to tell you if you don't take one.' And she wouldn't. She said she could stand whatever I had to tell her—no, that Jesus was her tranquilizer. She said those exact words . . . So she said, 'Well, what is it?' So I told her that there had been a plane crash and it could very well have been the flight that Ann was scheduled to take . . ."

Broder asks her to describe the week leading up to the funeral.

". . . and then on Wednesday, we were planning—I was thinking that if it had been true, that when we see her, I just hoped that she would be smiling. Because I hated to think that she was out there all by herself and nobody to help her, and all the agony she must have gone through. And if she called out for one of us, we weren't there, and they wouldn't let us come either. We had asked them if we could come [to Bluemont] and they said no.

"So, I was hoping that she had been smiling and that we could dress her in red and white, because those were her sorority colors, they were her favorites. Then we found out that the coffin was sealed. That we couldn't see her. So then there was this thing about whether or not to look. Because from the way Mother was talking, you know, I don't think the thought ever occurred to her what those bodies were like up there, and I didn't think of it either until they said the coffin couldn't be opened. And so, we couldn't dress her or any- thing. That thought was out. She was probably just naked in a zippered bag."

The courtroom is very still.

"So we tried to get Mother—we convinced her not to look. We went and got some flowers and the next day we went down to the mortician. And we had to talk Mother out of looking again. She still wanted to look. She said if it was just a hand or something and . . . if it was just a hand . . . I could have looked too. But I wouldn't have wanted to see a damaged hand . . . maybe, you know . . ."

Even Broder is silent.

"Saturday morning we got dressed for the funeral . . . Finally we proceeded to the morgue and we were in the car behind Mother, and when the hearse came out with the coffin in it, I could see Mother's head through the rear window of the car. I could see her head, you know, she couldn't even sit up straight. And so we went to the church. And Mother had really been trying. She had really been trying, but she wasn't very successful . . . When we got into the church, she said, 'I held it back as long as I could, but Lord, you know, I want my child. I want my child.' " Brenda looks across the courtroom. Essie English, hearing her own words, remembering the day, sobs. Sonia and Harold hold her hands. It is a long empty way between the witness stand and the bench where the family sits.

"I kept thinking that maybe—we didn't *see* her in there, maybe she wasn't on the plane. Everything is possible. There could be some explanation." Brenda turns to the jury, as though they might have it. "Maybe she wasn't going. Maybe we could find her. So, I tried to find her. I went into the house by myself and I made phone calls to all kinds of ridiculous places. And she wasn't there. She wasn't in any of those places. I couldn't stand the thought of not being able to find her. She must have been somewhere. There's always hope . . ."

"Do you," Broder asks, "carry that hope with you today?"

"Yes," says Brenda Rodriguez, "because I haven't found her. But there's always hope. When you lose hope, what else do you have? It is possible."

Broder looks at his legal pad. "I have nothing further."

"No cross-examination, Your Honor," says Booker.

"You may step down," Judge Bryan says.

Brenda walks back to her family. She and her mother hug.

"Call your next witness," the Judge tells Broder.

"Your Honor," he says, "with regard to the testimony, that's the plaintiff's case."

He does, however, wish to offer some exhibits into evidence. The Judge calls a bench conference. The matter of exhibits is settled. Broder asks the judge what "the time schedule is? It's five to five. My back is breaking, Judge."

"How much evidence do you have?" Bryan asks Booker.

"We have one economist who will take, I would assume, an hour," the TWA lawyer says.

"What's your economist going to testify to?"

"He's going to testify—"

"Has there been," Bryan suddenly asks, "any evidence of dependency in this case?"

"No, sir," says Booker, firmly.

"Why do you need an economist?" asks the judge.

Broder disagrees. "There's been evidence in the case of contribution, inclination to contribute and to help—"

"Mighty slim," says the judge.

Broder is in trouble. "The point is that this inclination or proclivity or propensity for the deceased to contribute—"

Bryan interrupts him again. "I will let you put the evidence on," he says to Booker. "I am still not sure I am going to submit the issue of dependency to the jury."

"If the Court please," Booker says, "if the issue of dependency has not been made out, I am certainly not going to put on an economist because nothing my economist can do is going to increase the concept of dependence."

"Dependence is a fancy word," Broder bristles. "The fact of the matter is that we have a child who was contributing to her sisters and brothers when they needed it . . ."

"I haven't ruled on it finally yet," says the Judge. "I just say I have got some questions about it. Certainly so far, as some of the beneficiaries are concerned, there's been no issue made as to dependence. Nearest thing you've got to it is Mrs. English, the mother."

Broder is in trouble. He's just heard the Judge say that half of his case may not have been proven.

But Bryan relents and tells Booker to put his economist on the stand. "I think there's at least enough evidence to go to the jury on Mrs. English's dependency."

Broder changes the subject. "With regard to the time and summation, I presume you are not putting any perimeters—"

"Indeed I am," Bryan cuts in.

"We have huge exhibits which—"

"How long," the judge curtly asks, "do you think you will argue?"

"I will require three hours," says Broder.

"I will give you an hour," Judge Bryan says. "That's more than I give most."

Plaintively Broder says, "I haven't read the exhibits either. It would take me—there are three statements supporting this young lady, indicating everything she said—"

"I have never given three hours summation in my life," Bryan says, "and in this case—"

"It's a matter of reading—"

"I have no idea of giving you any more than an hour."

"I need an hour and a half in my summation, and at least some time—"

"I have ruled on it, Mr. Broder," snaps the Judge. "You are lucky to get an hour."

"We think a half-hour would be ample time," Booker interjects innocently, regarding his own closing arguments.

Broder won't give up. "Would Your Honor allow me a half-hour to read my exhibits?"

"An hour total. You can spread it any way you want." He turns to Booker. "Call your next witness."

Thomas Kevin Fitzgerald, a senior economic analyst for the National Economic Research Association in New York—

TWA's answer to Professor Guilfoil—testifies that, generously computing Annie Killingsworth's contributions to her family to be one-third of her yearly disposable income, she would have given her mother and siblings a total of $41,568 throughout their lifetime.

Booker then has him describe how, to produce a yearly income of $1000 for Mrs. English (a figure he defended, when Broder objected to it, by saying, "I tried to establish from Mrs. English what amount she did receive. And I am saying in the interest of great liberality that the most she was receiving was a thousand dollars a year. I have taken the figure out of the air simply because I couldn't establish anything further from Mrs. English.") it would take capital presently valued at $13,907.

The jury is attentive, doubtlessly aware of the huge discrepancies between the figures mentioned by Broder and those elicited by TWA.

Booker asks his economist to explain the variance. Fitzgerald does so by saying that Guilfoil did not take into consideration Miss Killingsworth paying any income tax, nor her spending any money for her own survival. Guilfoil's total of $1,009,000 was simply, he says, a gross figure purporting to represent her expected, untaxed, unused earnings to the age of sixty-five.

Broder begins to argue with Fitzgerald while cross-examining him, is cautioned against that by the Judge. By 5:30, all the evidence in the case has been presented.

Admonishing the jury not to discuss the testimony or be influenced by any press accounts of the case, Judge Bryan adjourns the trial until ten o'clock the next morning.

The jury files out in front of Mrs. English and her children. In the corridor the insurance men are at work.

Tuesday, March 2, 1976, 10:20 A.M. Aaron Broder, in the same blue suit as yesterday, walks to the lectern facing the jury. "May it please Your Honor, my colleagues at the bar, Mr. Foreman, Ladies and Gentlemen of the jury. During the course of this trial, certain emphasis has been placed upon mathematical formulae, with regard to the evaluating of the damages sustained in this case. In attempting to arrive at a verdict concerning these five brothers and sisters, and this mother," he half turns to nod at Essie English, still in black, still huddled on the same bench—"the mathematical formulae which have been presented to you are indeed the least important part of the case, and refer to the smallest part of this case because, indeed, the very least of what has been lost to these brothers and sisters, and to this mother, is the loss of income from this decedent.

"And I say that, not because it isn't a considerable sum, but because, if indeed the total of one million ninety thousand dollars, or, one million nine thousand dollars of earnings were to be turned over to them, which of course it would not, that would still comprise only a minute portion of the total damages sustained . . .

"We are dealing here in this state now, in this enlightened day of 1976, with new laws and new ways of thinking. And a new way of thinking as incorporated in the law is to fully compensate, but completely compensate, persons who have lost loved ones through the wrongful act of others.

"And bear in mind at all times that when we come before you now and ask you for your verdict, this family doesn't come here with hat in hand." He pauses. "That's the vital thing to understand, that they don't come before you humbly—" his voice is loud now—"and pleadingly, charitably, asking you for compensation. Not at all! They come before you—with all due respect to you members of the jury, and we have the greatest respect for you—but they come before you saying, 'We are entitled under the law, in the State of Virginia, we are *entitled*, where there has been a wrongful death—and there was in this case—to full and complete compensation.'

"And that is the vital way in which I ask you members of the jury to look upon this case, just in the same way as if Trans World Airways, the defendant in this case, were to come before you in some commercial case where perhaps one of their planes had been lost or damaged and they came before you and said, 'We ask you for the full cost of our plane, thirty million dollars. We ask you for thirty million dollars for that plane, for every nut and bolt and every part of it, and every instrument within it and all the effort and all the work that went into building that plane. We ask you for full compensation for that plane, because someone wrongfully took it from us.' You wouldn't be evaluating that on sympathy for Trans World Airways. You would go back and say, 'Well, this corporation is entitled to this,' if indeed they were, as the family is in this court today.

"And so we ask you, likewise, to consider their rights. As His Honor has instructed you, 'not with anything you might have read or heard about in the newspapers, not with regard to other cases, not with regard to matters which you might have heard about—whether today or years gone by—but based upon an enlightened new law in this state which allows compensation for sorrow, mental anguish, solace, the loss of society, companionship, comfort, guidance, kindly offices, and advice of the decedent . . .'

"And you sit there, and you say, 'Oh, goodness, kindly offices? Nonsense. What kindly offices? How kindly offices? Why kindly offices?' And the reason for it is," Aaron Broder thunders on, "that some of us lay in the mud—lay in the mud to see that the laws in this country are protected and defended and remain enlightened."

The jurors seem slightly dazed. There are knowing looks among the attorneys in the spectators' section.

"And that's the important reason," Broder continues, "not because I say to you, 'They are entitled to the kindly offices of Annie Killingsworth,' not because I say, 'They are entitled to recover for companionship of Annie Killingsworth,' not because I say, 'They are entitled to recover for the loss of society of Annie Killingsworth,' but because, as I say, that some of us lay in the mud to defend those rights, and you listen to the law and see whether the law doesn't provide for this full cup of justice. Yes, for this full cup of justice, for this full measure of justice, to replenish the tears, the full cup of tears, of a mother and of a family of brothers and sisters.

"And then on the question of whether or not those tears should be replenished, I ask you to listen to the law again and see whether or not sorrow, yes, sorrow, not because I tell you that, a lawyer who comes here from New York, to represent a young lady who had lived in Ohio, and her family who lived in South Carolina, strangers before you, not pleading, not begging, not for charity—but for the upholding of the law of the State of Virginia where this crash occurred, and see whether or not that law provides what we are talking about, what the normal response might be." Broder stops, looks at his notes, collecting his thoughts and his breath.

"Yes, that sorrow, mental anguish, and solace—three words. Listen carefully, because they are recited along with all the law in this state to you, when His Honor recites them. And he doesn't stop. That's my job, as the advocate, to call your attention to those portions of the law that His Honor is going to refer to, so that you recognize why we are here and on what basis. So I ask you to carefully, carefully listen to His Honor, and listen to those words when he charges you relative to the law of this state and what applies in this state. And then say to yourselves, 'Oh, this lawyer from New York comes down and asks us for ten million dollars for that family. Well, maybe that's his view up in New York, and maybe that's what those folks do or would do, but this isn't what we'll do down here in Virginia.'

"What I say to you is this: You must put aside the past, the past has been put aside relative to the law in the State of Virginia, because this is new law, 1974 law. And this is the first case that's being tried of the Trans World Airways crash, so the eyes not only of the state but of the country are upon you and how your evaluation is, and what your evaluation is of new law and new concepts and the rights of people."

Broder then admonishes the jury to ignore any preconceptions they may have and decide the case on its evidence,

evidence he says, which shows Annie to have been "the star of their family—the masterpiece of their family . . . an extraordinary human being.

"It is difficult," he concedes, "to recapitulate a life in a short period of time, but the exigencies, realities, in terms of time, require that we do so. I would ask you to bear in mind that there are cases which go on—a partner of mine trying [a case] across the country, in San Francisco for weeks and weeks—" F. Lee Bailey is that same day summing up his defense of Patty Hearst—"and you folks have read about it in the press for weeks and weeks . . ."

"Couldn't resist bringing up Bailey, could he?" whispers a lawyer in the back of the courtroom.

Broder gets back on track, ponderously outlining Annie Killingsworth's outstanding performance in high school, in college, in her job with HUD. "In every single respect she exceeded, and she pulled herself up by the bootstraps to become the outstanding star, the masterpiece of this family. And masterpieces have gone for great sums of money. We know that. Paintings, millions . . . billions. Race horses . . . millions of dollars.

"Well the question is: is it so ridiculous—is it such a contemptible and despicable thing for these people to come in here and say to you, 'Yes we are to be fully compensated under this law, and so we ask you for ten million dollars?'

"And you sit back and you say, 'Oh, that's a ridiculous sum of money.' But the question is, is it ridiculous? Is it ridiculous in terms of what other values are placed on other things in this life? Can we not—on human suffering which is to be fully compensated—can we not see that they are fully compensated? Do we have a respect for the material, for the tangible things in life, or do we have a respect for those intangibles? And the point is, sure, intangibles are difficult, difficult of appraisal because the suffering is boundless. Not because it isn't there. But because it is so great. That's where the difficulty comes in.

"How many times have we heard," he asks, "when a
mother loses a child, how her gray hairs go down to the grave
in sorrow and in grief? We are talking about what's vital for
your consideration is the fact that this grief, this sorrow, this
anguish, this remembrance, this harrowing experience—the
most harrowing experience that I can conceive of a human
being—an air crash disaster—to be seared across your mind
as it has upon this family. Is that little girl of this mother lying in some plastic bag, dismembered? Do you recall those
thoughts in the minds of this family?"

The jury looks as though it would rather not.

"To open or not to open," Broder presses on, "to look or
not to look, to remember to the mother to think, even hopefully, maybe there's an arm there, maybe there's an arm of
my girl that I could look at.

"These are the things. Oh, we don't ask for your compas-
sion. We don't ask for your sympathy. That is not what is
involved. We ask for your objective appraisal of their com-
passion, of their sympathy, of their sorrow. That's what must
not be confused. Some of you might say, 'Oh, Mr. Broder's
appealing to sympathy.' I am not appealing to sympathy. I
stand up here proud to say it, that no human being can conceivably say if they understand what I am saying, that I am
not appealing to your sympathy at all. I am appealing to you
for your reason, to be applied to their sympathy, to their sorrow, to their grief, to their mental suffering, to their anguish,
to their harrowing experience. And for you to objectively look
upon it and say, there can be no doubt that they went through
this. *And going through their life is like looking through the
window of hell.*"

"And so the final question is," Broder, now drenched in
sweat, concludes, "what appraisal and how to appraise—and
each of you, in taking your sworn oaths as jurors here, was
asked by the Court, would black or white make any difference
in this courthouse? And your answer was no. And that applies,
of course, to human suffering, because it has never been

proven, not once has it been proven that the lowly born, a person of low station of life, of low economic income, or from some place that you never heard of in South Carolina, suffers any less or feels any less, tortures his mind any less, but on the contrary, perhaps more, when something extraordinary, when some precious gem that they had has been taken from them."

He tells the jury that he is going to save his additional time for his rebuttal and then he thanks them.

It is TWA's turn. Lewis Booker walks to the lectern. He lays his hands on it, looks at the judge, and back at the jury. "May it please the Court. No one within the sound of my voice today is immune from the tragedy of a sudden death, the tragedy of the death of a young person, and I doubt whether there is anyone in this courtroom today who hasn't been touched by that once or twice in his own life. So what do we do when that happens? That's really the question before us today. What can we do, after we have given our sympathy and our understanding to someone in that position? What more is there to be done?

"I have," he says, "the most difficult role one can conceive of in the courtroom, and that is to try to talk to you about the money which the family of Annie Killingsworth should receive. I wish I didn't have to do it. I know you wish you didn't have to decide that. But we do. We are here as officers of the Court to make that decision this morning.

"What I say to you is not evidence, as I pointed out in my opening statement. What I am going to say to you for the next few minutes is what I recall the evidence to be, and if I have misunderstood the evidence, or recalled it incorrectly, obviously, of course, it's up to you to trust your own recollection and not mine of the evidence. Nor shall I attempt to tell you what the law is. I believe I have an idea of some of the things Judge Bryan will tell you in a few minutes when he tells you that, and I will attempt to encompass them within my remarks to you this morning. But the evidence comes from the—the evidence comes from the witness, and the law comes from the judge, not from the attorneys.

"I told you yesterday that we were not going to talk about Annie Killingsworth to say she wasn't as capable as the evidence shows. You have heard the evidence here today. But that's not really the issue. The issue is not whether this was an exemplary young woman. All of the evidence is she was. The issue is what compensation her family shall receive. Whatever she was," he says with a touch of scorn, "she was not a race horse, and she was not a painting. I suggest it demeans her family to equate her to that." Booker allows himself a long silence.

"There are two aspects of the monetary recovery in this case. First, there is the question of loss of expected income. I believe Judge Bryan will tell you that you may award compensation for reasonably expected loss of income. Those words are extremely important—'for reasonably expected loss of income' which anybody has sustained. Now, that must be what one could reasonably have expected to receive—not by speculation, not by conjecture, but [by] what the evidence discloses to you the parties had a reason to believe they would receive.

"What income," he asks, "did the brothers and sisters reasonably expect they would receive from Miss Killingsworth? None. And none was needed. You have heard the brothers testify about their work, their employment. They are capable of working. They are working. You have heard her sisters testify—educated people, people seeking to better themselves, not people who are going to be dependent upon anyone. If there is one thing that comes through very clearly about what Miss Killingsworth tried to teach her brothers and sisters, it was to be independent. It was to stand on their own feet. It was to make their own way in life. They weren't looking to her for financial support. They didn't anticipate or expect to receive any income from her.

"Certainly, she gave them gifts from time to time, just as any of us would give gifts to our brothers and sisters or other members of our family. But that's gift. That's not something that's expected, something that's required, something that's needed. That's not the nature of the contributions you heard described on the witness stand. What you heard were gifts from a loving sister, not income to support someone who indeed did not need it. Mrs. English testified about receiving certain checks and sums of money from Miss Killingsworth. Again, that was not money Mrs. English needed to live on. She supports herself as a schoolteacher. Her husband is working, capable of supporting himself. They live comfortably, as Mrs. Wright told you, the first witness on the stand.

"So, Mrs. English does not look and did not look to Miss Killingsworth for financial support. She told you yesterday that she intended to work as a schoolteacher until she retired. Then she would have, of course, whatever pensions and social security available to her. She was not dependent and did not expect any income from Miss Killingsworth. She did testify as to receiving small amounts of money for repayment of a loan, or, if you choose to characterize it in some other way, as gifts to her. Mrs. English was very frank with you. She said she couldn't tell you how much that was—three hundred dollars a year, five hundred dollars a year. She simply couldn't make any statement at all. But there is no evidence that she expected Miss Killingsworth to support her the rest of her life, not at all. She's able to stand on her own two feet. I think that came out clearly in her testimony to you yesterday.

"What would Miss Killingsworth have done with her life? Who can say?" Booker asks, changing tacks. "Is there any reason to suspect that someone with her talent and ability wouldn't have become married, wouldn't have raised a family of her own, a family of her own that would have looked to her for support? Is there any reason to say that she was going to continue to work just for herself, and that her family in South Carolina would be looking to her for support? There is no evidence whatever to support the fact that any compensation for reasonably expected loss of income should be given. Remember," he tells the jury, "the burden of proof is on the plaintiff to show that there was a reasonably expected loss of income. It cannot be speculative. It must be reasonably expected. And what evidence is there of any such expectations in this case? I say to you there is none."

Booker then reviews the two opposing economists' testimony, dismissing Professor Guilfoil's million-dollar-plus figure —the plaintiffs' estimate of Miss Killingsworth's lifetime income—as "absolutely meaningless in the context of this lawsuit," and, reminding the jury that he had asked Dr. Fitzgerald "to assume that the family could reasonably expect to receive a thousand dollars a year, I suggest to you that that's more than the family ever has received, that there is no evidence that as much as a thousand dollars a year was ever received from Miss Killingsworth."

A few of the jurors quickly glance at Essie English. Holding her two eldest daughters' hands, she gazes at Mr. Broder's back.

"And yet," continues Booker, "I asked Dr. Fitzgerald to make that assumption. He made that assumption. He applied the discount rate suggested by the plaintiff's economist, Dr. Guilfoil. And he concluded that the sum of thirteen thousand five hundred dollars invested at the discount rate suggested by the economist for the plaintiff would yield that amount of money forever, and not touch the principal. And so, if you believe that anyone had any expectations, reasonable expectations—not speculations—of any kind to receive money from Annie Killingsworth, then it cannot have exceeded the amount that you heard testified to yesterday, thirteen thousand five hundred dollars, as sufficient to replace that sum."

He lets the figure register.

"The large element that has been mentioned to you by counsel for Mrs. English," Booker says, "the amount of ten million dollars, is really then for grief and suffering. But how," he asks, "can anyone put a dollar amount on grief and suffering. How can anyone say that one of the children should receive more or less than the others?

"Grief and suffering doesn't last forever. The loss of a child, the loss of a husband, the loss of a wife, that loss occurs. There is no way TWA or anyone else can put that life back here on earth. But the grief . . . doesn't last forever.

"Yesterday, you saw this family relive those tragic days at the time of the death and of the funeral. What one of us in this courtroom would not be reduced to overwhelming grief if we had to relive the death and the funeral of someone we loved? We all would be. We would all react just as Miss Killingsworth's family reacted, when we have to relive an event like that. But life goes on. Life must go on. And while we have all had tragedies in our lives, we must go on living for the present. And that grief does leave us. The memories don't leave us. The love and affection we had doesn't leave us, but the grief isn't going to go on forever. Mrs. English is still teaching school. Edgar is working at his job. Harold is working at his job. The sisters are continuing with their education. Life did not stop for this family when the death occurred. It was a tragedy. It was certainly a tragedy, a tragedy which caused overwhelming grief—grief which we suggest cannot be measured in any dollar amount, and grief which cannot and does not linger forever.

"I am not going to take much more time," Booker says quietly, "because I think I have said to you what I can say to you. Let me just summarize what I would like for you to remember when you retire to consider your verdict . . . . There are really two aspects of the dollars you will be asked to award. First, for reasonably expected loss of income. The evidence on that is clear. The economic testimony on that is clear. It's around thirteen thousand five hundred dollars, present value. And you must, of course, give present value.

"On the second amount, the question of how much to give this family for grief and solace and mental anguish, there are no dollars which can do that."

Broder is writing on a yellow legal pad.

"This family," Lewis Booker says, measuring every word, "doesn't want or need to be made wealthy because of the death of Miss Killingsworth. You have seen this family on the stand. No amount of dollars is going to replace Miss Killingsworth in their lives. There is no way one can suggest to you a figure.

"I am going to end today," he says, "as I did in my opening statement to you yesterday. Mr. Broder is not the only one who has lain in the mud for his country. He is not the only one who wants to see the laws enforced. But justice is a two-way street. Justice goes both to the family of Miss Killingsworth and to Trans World Airlines. So we ask you, when you retire to consider your verdict, that you be fair to both, not just fair to the family of Miss Killingsworth, not just fair to Trans World Airlines, but to both of them, because they are both here in this court, and they are both entitled to your deepest consideration."

Booker, who has been leaning over the lectern, stands upright. "And if you are fair to both of them in your own conscience, you can walk back in here to deliver your verdict and you can look in the eyes of anybody in this courtroom and you know you will have done your job. And more than that, no one can ask."

Broder rises for rebuttal. "I don't quite understand," he begins with undisguised sarcasm, "and I listened very carefully to my colleague who said to you in plain English, so that everyone in this court can hear him, politely, courteously, and as a fine gentleman which he is—that no amount of money can possibly compensate these people for the loss. That is the thrust of his final conclusion." He shakes his head. "And the other part of his conclusion is that it's impossible to measure grief. It's impossible to measure anguish. So you give nothing for it ... And the third item was that there are others who have lost loved ones.

"Let's deal with these thoughts that counsel has thrown

out to you and, allow me if you will, to first put aside this question of other matters, other cases, other evaluations. In trying to measure the grief of an individual—if you have a confirmed drunkard, some vile individual and let's say it the way the kids do, say it as it is, sometimes the grief for those who have lost that individual might be poignant but not quite so poignant as in other matters and in other losses of individuals. If you have a very small family with one individual affected, one person's grief may be immense but, again, it's one person's grief that's left in that family. Certainly to be compensated, certainly poignant . . . If you have an older person, in his eighties, and his life is taken, surely there is some grief. However, there again, it may not be in the same ballpark with what happens when a parent loses a child, when a brother loses a sister, who is the bright, beautiful individual with everything to hope for and to look forward to. And when the grave and a wrongful death robs that individual of that life, then reflect upon what it means to those who loved that person.

"And when [my colleague] talks in terms of the bloodless accounting, and it is a bloodless accounting—the double-entry bookkeeper kind of thing—that goes for the specifics, when someone comes into court with a case, as they say, we had a contract, and this corporation says we want it—that's double-entry bookkeeping. That's bloodless accounting. And that's proper for that. But that has nothing to do with the aspect of grief and sorrow, and the loss of companionship that we are talking about. That's in this case over and over from the defense—to throw up into the air and to counterflank the true issue. The true issue in this case is what is the grief and the suffering and the mental anguish?"

Broder then attacks Booker's contention that Annie Killingsworth had not, in fact, ever supported or intended to support her family. "You can," he says, "ignore that nonsense that he made reference to in terms of what they could expect. I would expect a girl like this, who could conceivably rise to all stations in life, to give her family everything. And to come back down there one day and provide everything for her mother, and for the brothers and sisters who didn't make it. So the pecuniary loss itself in this case is extremely substantial. I am not going to put a figure on it, because that's the least important matter, and I don't want to confuse the issues here.

"We talk of individuals in the case. We talk of Sonia, who was closest to Ann, who was so tragically affected by her death. Sonia, the introvert in the family; Ann, the extrovert, who helped her out of her shell, Ann, who gave her the confidence to face existence with those who may have had an advantage of social status . . . You will recall, [she had] a feeling of inadequacy with the upper classes of life. Sonia, who was helped so much by her, Sonia who grieves so great. You heard her, these tears in this courtroom were real tears. They weren't fake. They were real from every member. Every sob and tear that you have seen here is only a manifestation of something which has been repeated over and over again . . .

"Ann, who taught her about the social graces and bought makeup and clothing and gave her more and more material gifts. More important was the giving of love and the understanding and confidence. This is what she spoke about. And Brenda, Ann's lifelong . . . so big to Brenda . . . was almost impossible—" he searches through the pages of yellow legal paper in front of him . . . "just some notes I made of the contents of the testimony—to believe that she had been taken from her. Efforts in demonstration that there was hope that she was still alive . . . the mother collapsing at the cemetery and at the church, the frantic efforts to find out, to hope that she was alive. All that is compensable.

"And Essie English, the mother. Can we talk about the emptiness in her life? Is there anybody who knows someone who has lost a child who can ever talk about it to a stranger and never hear a word from her? It's all internal grief, and it's there, and you know and I know with that person, that person lives every minute of his or her life—of a parent who has lost a child—of a parent who has lost a child . . . I heard people say, and I am sure that you have heard them say that no greater, no greater loss, no greater grief, no greater cancer, no greater heart attack—nothing can befall a person greater, more disaster to a person's life than to lose a child and to remain, almost like a freak of human nature, living while that child is dead. These are the thoughts that go on in a parent's mind.

"Edgar, that she referred to as 'Turk.' Who was jealous of his sister's greatness and who betrayed a certain kind of guilt feeling—the big brother who loves his sister's singing, and the singing brought him happiness and comfort and assurance. And whose singing he had never heard better nor had the members of the family. And this great big guy sitting there reduced to tears.

"And Dorcas, who thought of her sister Ann as the prettiest sister, who made her feel special and who still was expecting her sister to call her, even to this day. Consider the effects of life upon this fifteen-year-old girl, Dorcas.

"And Harold English, that was the young man who seemed the sturdiest in the whole family, but who spoke of having been made a stage performer. And when Ann said to him, 'Why don't you go on the stage?' he laughed. He said, 'How can it be me, in me. Big old me?' And she says, 'Go ahead.' And sure enough, he went on and he enjoyed life a little bit more because of her. And he said it wasn't fair for Ann to die.

"So these are all of the elements. We have prepared it and presented to you within a day," Broder finally says. "Then we call upon you for your evaluation, for you to stand up proudly and announce before this state, before this country, what the view of Virginia is upon this new law, what your evaluation is of human life, human suffering, and of the loss that has befallen this family."

One last deep breath. "On their behalf, I want to express

to you our grateful appreciation of your angelic patience with us during the course of this trial. Thank you."

He walks back to his table, wet, breathing heavily, oblivious to his clients who sit there behind him in a long mournful row.

Judge Bryan calls a short recess.

Out in the hall, other plaintiffs' attorneys congratulate Broder on his closing arguments—the same men who yesterday privately scorned his tactics, his trial of tears.

"Finest argument I've ever heard," says one.

"Hope the six members of the jury agree." Broder is confident, gracious.

The lawyers drift off into corners and guess at the jury's eventual award. "Essie English saved his case," says one. "Yeah—until she was cross-examined." "No," says another. "Brenda Rodriguez did it." "Two hundred?" "If he's lucky . . ."

Overhearing the speculation, Fred Schaffhausen, the insurance man, smiles. On the mere strength of Lewis Booker's defense, he's been settling cases all day.

The lawyers wander back into the courtroom. A huge black security guard watches them. "You know," he says, "I don't much care for these civil cases. It's just all these innocent people."

Reading from notes, Judge Bryan charges the jury: "Now that you have heard the evidence in the case and counsel's arguments, the time has come to instruct you as to the law governing the case. You, as jurors, are the sole judges of the facts and, therefore, what the Court or counsel may say to you regarding the evidence, or their recollections of the evidence, is not binding on you. You are, however, duty bound to follow the law as stated in the instructions and to apply that law to the facts as you find them from the evidence before you.

"You are not to single out one instruction alone and give

it special attention, but should consider the instructions as a whole. Neither are you to be concerned with the wisdom of any rule of law. Regardless of what you might feel the law to be, it would be a violation of your sworn duty as a juror to base a verdict upon any other view of the law than that given in the instructions. You have been chosen and sworn as jurors in this case to try the issue of the fact presented by the allegations of the plaintiff, that is, the estate of Annie Mozelle Killingsworth on the one hand and the response to those allegations filed by Trans World Airlines on the other hand.

"You are to perform this duty without bias or prejudice as to either party. The law does not permit jurors to be governed by sympathy, prejudice, or public opinion. The parties and the public expect that you will carefully and impartially consider all the evidence, follow the law as stated by the Court, and reach a just verdict, regardless of the consequences.

"Statements and arguments of counsel are not evidence in the case, unless made as an admission or stipulation of fact. For example, there has been referred to in the closing arguments the amount sued for, or the suggested amount that you are to return as your verdict. That is not evidence in the case. It is merely the statement or argument of counsel, and since it is not evidence in the case, it should not be used by you or considered by you in arriving at your verdict.

He tells them that the evidence consists solely of the sworn testimony of the witnesses and of all the exhibits which have been admitted into evidence, but that "in your consideration of that evidence, you are not limited to the bald statements of the witnesses. On the contrary," the Judge admonishes them, "you are permitted to draw from the facts you find to have been proved such reasonable inferences as seem justified in the light of your own experience. You, as jurors, are the sole judges of the weight to be given the evidence and of the credibility of the witnesses, by which is meant their worthiness of belief—and, in ascertaining the credibility of the witness on the witness stand, his apparent candor or fairness, his bias if any, his intelligence, his interest or lack of it for knowing the truth and having observed the facts to which he has testified.

"A witness is presumed to speak the truth, but this presumption may be outweighed in the manner in which the witness testifies, by the character of the testimony given, or by contradictory evidence." Bryan pauses briefly. A U.S. marshal stands by the now-locked courtroom door, locked to avoid the jury's being distracted by comings and goings during the judge's charge.

"As indicated earlier," Bryan resumes, "the sole issue in this case for you to decide is the issue of damages. And the issue of fault need not concern you. TWA, Trans World Airlines, has accepted full responsibility for the payment of all damages which you award. They are not to be penalized by this or for this. In other words, your award should not be increased because of this. By the same token, the plaintiff is not to be—the decedent's estate is not to be—penalized by this. . . ."

Judge Bryan tells them that damages are not presumed and cannot be based on speculation, that the burden of proof is on the plaintiff, but that the alleged damages need not be proved "with mathematical precision," or to exist "beyond a reasonable doubt, or such a degree of proof as produces absolute certainty, since in human affairs absolute certainty is seldom possible."

The damages to be awarded, he says, are to be "those that seem fair and just to you and shall include, but may not be limited to, damages for the following: first, sorrow, mental anguish and solace—which may include society, companionship, comfort, guidance, kindly offices and advice of the decedent; second, compensation for reasonably expected loss of income of the decedent; third, compensation for reasonably expected loss of service, protection, care and assistance provided by the decedent; and, fourth, the reasonable funeral expenses," which, he notes, amounted to $760.

Bryan adds that the jury may apportion the damages

among the family as they deem fair and just, and ends by reminding them that though their verdict must be unanimous,
"you should not, in the course of [your] deliberations hesitate
to re-examine your own views and change your opinion if
convinced it is an erroneous opinion. Remember at all times,"
he says looking up and at them, "you are not partisans. You
are judges, judges of the facts. Your sole interest is to ascertain the truth from the evidence you have heard here in court."

He motions for the two opposing attorneys to approach
the Bench. "Any objections to the Court's charge."

Broder says no, though he has one request.

The judge remembers. "I forgot to tell them about Miss
Killingsworth. I will tell them."

Booker objects to a number of things and is overruled.
Bryan once again turns to the jury. "I have been advised and
I don't know that it is an important matter to you, but in
case there is any misconception, that Miss Killingsworth was
unmarried. One of the witnesses apparently referred to her as
Mrs. Counsel was concerned that you might have attached
some importance to that in one way or the other. She was
unmarried at the time of her death."

This is—apart from the fact that Annie's last name was
different from that of all her family's—the only indication the
jury has been allowed to have that there may have been a man
in her life.

The Judge tells them that they may now retire to consider
their verdict. It is five minutes past noon.

The courtroom empties slowly, the lawyers leaving their
briefs, their books and papers on their tables. By 2:30 they
begin to gather again, after long meals, in the halls of the
courthouse. The jurors, having deliberated for an hour before
lunch, and taken an hour off to eat, are back in the locked
jury room.

Broder chats amiably with the handful of local reporters
covering the trial. He says settling a case "is a game," but
going to court, to trial is "an art." Obviously hopeful that the

Killingsworth trial will prove to be a landmark jury award—
the first under the new law—he talks disparagingly of attorneys who, "because they are scared to death to go into court"
settle early for an easy buck.

At 4:15 p.m., there is loud knocking against the inside of
the jury room's door. A U.S. marshal pokes his head in. A
verdict has been reached. The courtroom fills rapidly. Annie
Killingsworth's family look at the jurors filing in.

Ceremoniously the clerk of the court asks, "Ladies and
gentlemen of the jury, have you reached a verdict?"

A man in a red shirt, the foreman, says, "We have."

The clerk asks for it. It comes in an envelope. The clerk
opens it and reads: "Civil action 381-75-A, Jackson Rose, Administrator of the estate of Annie Mozelle Killingsworth, versus Trans World Airlines, Incorporated. We the jury find in
favor of the plaintiff in the amount of $90,760 . . ."

Broder lowers his head into his hands.

". . . and apportion on the verdict as follows: Wright's
Funeral Home, for funeral expenses, $760; to Essie R. English,
mother, $40,000; to Harold Tobin English, brother, $10,000;
to Sonia English, sister, $10,000; to Dorcas English, sister,
$10,000; to Brenda Rodriguez, sister, $10,000; to Edgar Tobin, brother, $10,000. Total verdict, $90,760. Signed: Donald
R. Owen, foreman, dated March 2, 1976."

The clerk looks up. "Ladies and gentlemen of the jury, is
this your unanimous verdict?"

"Yes," they say in quiet unison.

Broder requests that the jury be polled.

"As your name is called, will you answer in the affirmative
if this is your individual as well as your collective verdict,"
Judge Bryan asks.

The clerk reads off the jurors' names.

"Marie M. Neri?"

"Yes."

"Donald Owen?"

"Yes."

"Marie Palmer?"

"Yes."

At each "Yes," the heads of the Englishes sink a little lower. Only Edgar Tobin, the older brother, looks up, glaring at the three men and three women who had just chosen not to make him or his family rich.

"Sadie Prattie?"

"Yes."

"John Reddick?"

"Yes."

"Your Honor, he didn't call me—Maurice Norton."

"Mr. Norton," the Judge asks, "is it your individual verdict?"

"Yes."

Judge Bryan nods. "The jury is released and discharged," he says, "and the Court wants to thank you for your service. The Court will stay in session while the jury leaves the room."

The jurors file out between Broder and his clients.

"Any motions?" asks the Judge.

"Yes, Your Honor," Broder says. "I respectfully move to set the verdict aside on the grounds that it is against the law, against the facts, and entirely inadequate to cover the damages that have been sustained by this family—by the sisters and brothers, as well as by the mother in this case."

"The motion will be denied," Bryan says flatly. "Judgment will be entered on the verdict."

The trial is over.

The courtroom comes to its feet as the Judge sweeps back to his chambers. Hesitantly Broder turns toward Essie English and her children. They watch him coming.

He motions them to a corner, by a window. The discussion is brief. They leave.

Broder gathers his papers. Two reporters accost him. "Well?"

"It's obviously far better for TWA to kill its passengers,"

he says bitterly, "than it is for them to maim them."

"Uh . . . will you appeal?"

"I doubt it," Aaron Broder says. "After her lawyer gets his third, Essie English only has twenty-seven thousand dollars. There won't be any money left."

DOLLARS AND CENTS

December 1, 1976 . . . Up on the Blue Ridge of Virginia only the trees, now cracked and winter-stripped, bear truncated witness to the passage of TWA Flight 514. This is a tough month, as one woman widowed by the crash says, but at least the accounts are settled and the books closed on this, the second, anniversary of death. Almost.

Thirty-two of the eighty-five claims against TWA have been settled without the threat of lawsuits, and thirty-nine more *cases*, as the insurers call them, have been closed (with settlements to the families of the victims ranging from one thousand to nearly one million dollars), after the filing of Wrongful Death suits, but before—and in some cases only minutes before—going to trial. Six cases did get to juries. The awards ranged from $36,000 to $500,000. Eight cases are pending.

And the airplane, what was left of it—26 tons of titanium mixture—was sold for scrap at $350 a ton.

"It wasn't good scrap," the dealer said. "It wasn't pure."

EPILOGUE

"They were somebody to somebody." The thought was so obvious that I kept it unvoiced. There were cheers at the bar anyway—a lucky touchdown on television. The crash was three hours old, an interruption in the football game.

Late that night—a Sunday night in a cold beach house with what was left of the crossword puzzle and a smoky fire—the names were made public, the roll call of the unknown on a flickering screen. The news was that they had died, instantly, whoever they were, all ninety-two. The television anchormen harped on the number. For them there was not much else. Then they'd strain a little, mumble that a lady FBI agent had been aboard, a couple of congresssional aides.

"... somebody to somebody." The search began, spurred by a surprisingly personal—odd because I hadn't known any of them—sense of wrong, a refusal to accept the outright, cold, final dismissal of so many lives. Valuable lives, I hoped to show, important.

The search began—a drive to Washington, D.C., for the investigations, a drive to the midwest. Phone calls, halting and awkward. Reactions: bewilderment, anger, suspicion, acceptance.

Then, slowly, on other trips, the voices lost their mistrust, became comfortable. There were tears and eulogies, deliberate fabrications, and a truth all its own. More tears. Wrenching searches for words. Retreats to the language of soap operas. Euphemisms. More tears. Nightmares described, then shared.

I dreamt their dreams and woke in towns on the map.

Cross-country drives with radio stations going off the air, knocking on doors seeking directions, direction. In trying to measure how far the ripples spread, to fathom their loss, I saw that for those involved it was immeasurable, yet even for them, survivable. It had to be.

That their mother, their wife, their son, daughter, lover died in a plane crash was, admittedly, different. Not like a wreck on the highway or cancer, or even combat. A plane crash, because somehow public, becomes a happening, like murder on a crowded avenue. Yet, regardless of how they died, for those left behind, the gut-tearing emptiness, the mute three-in-the-morning despair is the same—an intimation of their own mortality.

Those ninety-two were not our friends, our family. Though we can empathize, even seek to lobby for change, the loss is that of those involved. The fabric of society was not rent, just rumpled. We pay lip-service respects, the bureaucracies shuffle papers, checks are made out and endorsed, flowers cut and laid by gravestones.

It ought to matter.